Extrasolar Planets

Extrasolar Planets

A Catalog of Discoveries in Other Star Systems

TERRY L. KEPNER

McFarland & Company, Inc., Publishers
Jefferson, North Carolina, and London

LIBRARY OF CONGRESS CATALOGUING-IN-PUBLICATION DATA

Kepner, Terry L.
　　Extrasolar planets : a catalog of discoveries in other star systems / Terry L. Kepner.
　　　　p.　　cm.
　　Includes bibliographical references and index.

　　ISBN 0-7864-2405-2 (softcover : 50# alkaline paper)

　　1. Extrasolar planets—Catalogs.　I. Title.
QB820.K47　2005
523.2'4'0216—dc22　　　　　　　　　　　　　　　2005024652

British Library cataloguing data are available

©2005 Terry L. Kepner. All rights reserved

No part of this book may be reproduced or transmitted in any form or by any means, electronic or mechanical, including photocopying or recording, or by any information storage and retrieval system, without permission in writing from the publisher.

Cover art: NASA/JPL-Caltech/R. Hurt (SSC)

Manufactured in the United States of America

McFarland & Company, Inc., Publishers
　Box 611, Jefferson, North Carolina 28640
　www.mcfarlandpub.com

Contents

Preface	1
Introduction	5
How to Use This Book	7
Planetary Discoveries	35
Appendix A: Withdrawn Discoveries	171
Appendix B: Tentative Discoveries	173
Appendix C: The Constellations	177
Appendix D: Masses and Orbital Characteristics of Extrasolar Planets	180
References	187
Bibliography	191
Index	195

Preface

This book came about because I wanted to write a science fiction story that was reasonably accurate in its descriptions of other star systems. Unfortunately, I discovered, finding out such information is difficult. While websites list information about extrasolar planets, they don't really tell you that much about the primary. And the sites that discuss the stars themselves don't give a lot of information about the planets that orbit them.

Fortunately, my lifelong love of astronomy, and space in general, was a tremendous help. My major in astronomy (University of Arizona, Tucson) certainly helped as well.

Thus began my quest. After a year, I had most of the information I needed, but then I thought other writers might be interested in what I had found. That meant documenting what I had discovered and making it a bit more formal than the series of file folders I had accumulated.

Some websites put the planets in order of how far from the primary the planet orbits. That works until a new planet is discovered and you have to reorder your list—easy to do on a website, or in a file folder system, not so easy on paper. Alphabetical order and numerical order suffer from the same handicap, so I decided that the only order that makes any sense is by date of discovery.

I immediately ran into a problem with that. There is no single location where planets are announced. Sometimes different groups announce planet discoveries almost simultaneously (literally on the same day and without any knowledge of the other group). Plus, data suggesting a planet discovery may take several years before it is convincing enough to actually publish the information, but the astronomers may discuss the "planet" with other groups, leading to it being mentioned as a "discovery" in a paper before it has been announced by the discoverers.

So, do I list the planets by when the paper announcing the discovery was published? I decided the only logical method was by announcement date wherever possible, then by submission to publication date (several papers announcing a discovery took over a year to be published after they had been submitted to a journal), and finally by publication date when the other two are not available.

This leads to some interesting discrepancies between various "discovery" dates listed on websites. For example, several multiple-planet discoveries were announced in one paper, but one or more of the planets were announced previously at a symposium or press conference.

Because I intended this book as a tool for people who are not astronomy or physics graduates, I tried to make the book as math-free as possible. The explanations of star life cycles, for example, is necessarily simplistic, but not too much so. On the other hand, I tried to include as much information about each star system as I could. Surprisingly, we know very little about many of the stars we are finding to harbor planets.

One thing we have discovered is that the metallicity ratio (the ratio of iron to hydrogen in a star's atmosphere) is an important indicator as to whether a star has planets. Using our sun as the standard, more than 80 percent of the discovered planets orbit stars with a metallicity the same as or higher than ours. Why this is so is the subject of much debate, but no conclusions have been reached. Similarly, the discovery of so many "hot" Jupiters (giant planets orbiting inside the orbit of Venus) has thrown all previous theories of planet formation out the window. New theories are being formulated, but so far none are adequate to explain both the metallicity and hot Jupiters.

On the other hand, the discovery of so many planets in such a short period of study gives hope that there might be habitable planets out there as well. It is unfortunate that our instruments can't detect the planets we most want to find. An Earth-sized planet exerts a pull on its primary as it orbits, making the primary wobble back and forth about two to three meters per second. Our current instruments have a lower detection threshold of about three to five meters per second. Plus, most stars that are like our sun have a "noise level" or "jitter" (sunspots) of about three to five meters per second.

Better instruments are being designed, and several space telescopes specifically for planet hunting are being funded.

The data are as accurate as I can determine. Astronomy is not an exact science in distance measurement. As a result, a lot of the figures regarding planet size and mass, orbital distance, and even star mass, are in the range of "we think this is right, but we could be wrong." Thus, different papers will list different numbers for the mass and orbital parameters of the "discovered" planets based on different estimates of the primary and its distance. Similarly, websites will list different numbers for the planet statistics.

The references in Appendix E are a comprehensive listing of the papers I used in putting this book together. Most are available for download in the Astrophysical Journal web archive, astro.ph. (http://arxiv.org/archive/astro-ph).

Finally, I hope you find this book as useful as I do.

Terry Kepner

Introduction

For many years astronomers looked in vain for evidence of worlds orbiting other stars. Until 1989 that remained an elusive goal. That year D.W. Latham, et al.,[1]* were studying HD 114762, measuring its rotation using a method known as radial velocity measurements.

They were measuring the Doppler shift caused by the star's rotation. The side spinning toward Earth has a slight shift of spectrum toward the blue, and the side rotating away from Earth has a slight spectrum shift toward the red. These measurements are on top of the base Doppler shift the star has in its movement through space in reference to Earth.

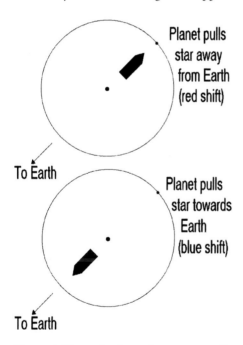

Figure 1. Measuring how the spectrum of a star shifts according to how the star moves when pulled by a planet in orbit.

After several months of measurements, they noticed their data points (individual measurements of speed at a particular time) had a "drift" that was repetitive and regular. They concluded that the only reasonable explanation was that the star was being "pulled" back and forth by an orbiting body. Thinking they had discovered a small stellar companion to HD 114762, they carefully, and

---

*Superscripts refer to papers listed in References in the back of the book.

unsuccessfully, searched pictures of the star for the telltale signs of a dim star. Going back to their data, they concluded they were seeing evidence of a giant planet.

Careful analysis of their measurements led them to the conclusion that HD 114762 is orbited by an object that could be as small as nine times the size of Jupiter. Unfortunately, we do not know the inclination of the object's orbit in relationship to Earth. If the orbit is edge-on to us, then the object is definitely a planet. As the angle of the orbit, as seen from Earth, goes up, so does the mass of the object. The mass of the object could go as high as 80 times the size of Jupiter. Because the theoretical limit for an object to begin "burning" hydrogen (actually deuterium, a hydrogen atom with a proton and a neutron in the nucleus) is for the object to be 13 to 15 times the size of Jupiter, the object orbiting HD 114762 could really be a brown dwarf.

This was still the first time an object that small had ever been detected. It was several years before anyone else repeated their discovery with another star, 51 Pegasus,[2] in 1996.

Over the next few years, other stars were discovered with planetary companions. There was some disagreement over whether these were planets or brown dwarfs, especially because all these detections involved gas giant planets that were far closer to their primary than any theory of planet formation had predicted. It was only when HD 209458, a planetary candidate, was detected passing in front of its primary that the radial velocity measurements were proven correct in their interpretation of these objects as planets, not permanent sunspots or brown dwarfs in highly inclined orbits.

As of August 1, 2005, there are about 161 planetary candidates in 135 star systems, ranging from 8.25 light-years to 5,000 light-years from Earth. Most have been detected by radial velocity measurements. One was detected by visual means (plotting the primary's wobble as it moves through space and is pulled from a straight line by the orbiting planet's gravity), seven others by their transit in front of their primary.

At first the planets discovered were close to their stars, but this was because it is easier to detect and confirm a giant planet when it makes several orbits during your observation period. A planet that takes 70 years to complete one orbit is not going to give a complete picture in the radial velocity measurements when one has been taking measurements for only three years.

Several teams are in the planet search effort today, using different techniques and equipment.

How to Use This Book

Accuracies and Estimates

Please be advised: While most stars listed are close enough for direct measurement of their distances, tiny differences in the measurements taken by astronomers can give very different answers for those distances. Minute variations in the temperature of the equipment, the accuracy of the equipment, and atmospheric disturbances can affect the readings. Do not take the numbers listed here, or anywhere else, as gospel — everything is at least plus or minus 10 percent (and in some cases plus 100 percent or minus 50 percent).

The secondary information about the stars is also subject to interpretation. For example, there are several theories for determining the size of a star we see only as a pinpoint. The diameter of Alpha Bootis, also called Arcturus, has been measured as 22 solar diameters, 29 solar diameters, 25 solar diameters, and 30 solar diameters. With a solar diameter of 870,000 miles, this gives a variation for Arcturus' diameter of from 19 million miles to 26 million miles.

Other uncertain factors relate to binary systems. When two stars are close together, we can photograph them periodically and put together a total picture of how the two move together through space. However, there are limits.

If the two stars are too close, we cannot see them. Fortunately, with the use of a prism and a camera attached to the telescope, we can differentiate the stars and determine when one is closer than the other. This is especially true when one passes directly in front of the other. These stars are called spectroscopic binaries. While we cannot observe them directly, we know they are there. The limiting case is when we are looking directly down on two stars. We know they are orbiting, but that is all we know.

If the stars are too far apart, we have another problem. We have not taken enough pictures to determine an orbital path, the time it takes the two to complete one orbit. Because glass-plate astronomy is only 100 years old, and only recently have we photographed the whole sky, orbital information is quite shaky on many systems. If we have seen only 50 years of what we think is a 10,000 year orbit, it is just as likely that the period could be 8,000 or 12,000. Only binary stars with orbits under approximately 100 years have orbital periods that are known with any kind of accuracy. Even those can be off by up to 20 percent. The other binaries could be off by as much as a factor of 2 (minus 50 percent to plus 100 percent) or more.

Other bits of data are also of dubious accuracy.

So, in examining the diagrams and their relevant data, one must remember that much of the information falls into the "best guess" category. Any claims that other data are better should be required to be proven. Professional astronomers can and do admit to errors.

Because this book is a summary, the explanations are necessarily simplistic, and are intended only to provide a sense of what is happening (at least according to commonly accepted theories). For a list of abbreviations and definitions used in this book, see the "Definitions and Measurements" later in this section.

The information herein comes from hundreds of published papers in *The Astrophysical Journal*, *The Astronomy and Astrophysical Journal*, *MNRAS*, the California & Carniegie Planet Search website, the Extra-solar Planets Catalog, SIMBAD, and several other print and web sources. (These websites mostly agree in their information, but may disagree.)

For some interesting reading on peculiar and unusual stars, read *Stars and Their Spectra*.[85] This book breaks the stars into their different spectral classes and discusses the average and uncommon types found. Little math is used, the author concentrating, instead, on explaining what the results of the math mean in real terms.

A writer worried about developing a believable space-traveling society should consult *Gurps Space*, a fantastic resource from Steve Jackson Games. Originally developed for the role-playing games market, this book provides well-developed research on types of technology, gadgets, weapons, environments, starships, space combat, and planetary civilizations one could reasonably expect to find in space. It even discusses different types of starship engines (how they might work) and the pros and cons of each.

Stellar Evolution

The accepted explanation for star formation is that a cloud of gas and dust several light-years in size (light-year: the distance light will travel in a year at 186,000 miles per second) develops an eddy or swirl. (The term "cloud" is used advisedly. In practicality it means several molecules per cubic meter instead of per cubic kilometer. The middle of a dense stellar dust cloud is still purer than the best vacuum scientists can construct on Earth.) As centuries pass, a swirl becomes stable and its movement develops a gravitational core. As more centuries pass, the swirl becomes thicker and pulls more gas and dust into its pattern. Smaller swirls may be dragged in, to be ripped apart by tides or to become subeddies orbiting the primary one.

Its mass increasing, the swirl begins to collapse as the dust particles exert minute gravitational pressures on each other. More time passes. How much? It could be centuries; it could be millennia.

The gas and dust particles begin to collide with one another at the center of the swirl, causing heat. As the swirl continues to contract, its core gets denser and hotter.

Gradually, as more centuries pass, the swirl captures a mass of dust and gas that can range from several times that of our sun to thousands of times larger. Its core gets denser and hotter.

Protostars: The Next Step

Eventually, a protostar takes shape. This object gives off a great deal of heat, but all of it is the result of collisions between its component particles forced together by gravity. The heat generated would vaporize the entire Earth and the pressure approaches that of the atmosphere of Jupiter. But there is nothing yet at the core to resist the crush of gravity. The density and heat are high enough so that the core is filled only with hydrogen atoms, the smallest existing atoms.

Astronomers have located what they think are protostars in some large dust clouds. The protostars are several times larger than our solar system. There are doubtless many subswirls and eddies buried in them. The protostar has a definite surface and atmosphere, but a thick dust cloud conceals it from direct view. Protostars are known only by their heat signature in the deeper mass of the dust cloud hiding them.

Eventually, if the stellar mass is more than 13 times the mass of Jupiter, the heat and pressure reach a point where a deuterium atom (a hydrogen atom that has a proton *and* a neutron as its nucleus) approaches close enough to a hydrogen atom to fuse into helium-3 (releasing a gamma ray). This generates a tremendous amount of heat. If the stellar mass is less than 80 to 100 times the mass of Jupiter, this is the stopping point. The star, a brown dwarf, burns through its supply of deuterium in a short time — between a few million years and 20 million years. It then spends the rest of its existence cooling off. This will take many billions of years.

If the stellar mass is greater than the brown dwarf maximum, the pressure and heat rise to the point (5 million degrees Kelvin*) where two hydrogen atoms overcome their mutual repulsion, fusing to form deuterium (releasing energy as one proton becomes a neutron). Then the deuterium atom fuses with another hydrogen atom to form helium-3 (releasing a gamma ray). The helium-3 fuses with another helium-3 atom, forming helium-4 and two protons. This chain reaction releases a great deal more heat, many thousands of times more than is generated by gravity-induced collisions. The atoms surrounding the new helium atom are forced away from it by this heat. For an instant the pressure drops far below the point where hydrogen will fuse. Then gravity reasserts its dominance and forces the atoms back close together. The newly formed helium atom is trapped under the gravitational pressure of all the hydrogen atoms above it.

More and more hydrogen atoms begin to fuse using this proton-proton exchange cycle. Each time this happens, a great deal of heat and pressure are exerted, pushing the other atoms away and slowing the collapse of the protostar. Gravity is still more powerful, and the protostar continues to condense and collapse.

Star Birth: The Delicate Dance

The trigger point of star formation depends on the mass of the collapsing protostar. At some point, the number of fusing hydrogen atoms is so high that their explosive pressure overcomes gravity and stops the gravitational collapse of the protostar. Not enough energy is released in these fusions to blow the protostar apart. Gravity is still a tremendous force. When the gravitational collapse stops, the pressure falls off, the atoms no longer fuse, the heat decreases, and gravity takes over and crushes everything together again.

---

*Astronomers use the Kelvin temperature scale, which has as its starting point absolute zero. Centigrade (also called Celsius), the more common temperature scale used in daily life, has as its starting point the freezing point of water. The difference between the two is that the freezing point of water is 273° above absolute zero. To convert a Kelvin temperature to Centigrade, simply subtract 273. The formula for converting Celsius to Fahrenheit is: °F = 32 + (9/5 × °C).

Now a delicate dance begins: The hydrogen atoms fuse and make everything move farther apart. Fusion slows. Then gravity forces everything back together. The result is a mix of hydrogen atoms fusing into helium, slowly filling the star's core with helium. The heat generated is immense. When this heat finally works its way to the protostar's surface it blows away, like a great wind, most of the gases and dust in its vicinity. The protostar is now a real star. It has a mass from one-tenth to 100 times the mass of our sun and a surface temperature from 3,000 degrees to 50,000 degrees Kelvin.

The time scale for this process is not really known, but some general statements can be made. More than 3,000 years ago, the small star cloud Pleiades was named by the ancient Greeks as the Seven Sisters. People could see that the cloud contained seven bright stars. Today, on a clear night, it is not difficult to see eight stars in the Pleiades. A small pair of binoculars reveals 22 stars. Several new stars have been born in the last few thousand years.

Really Big Stars Won't Have Inhabited Planets

Most astronomers agree that planets form within 10,000 years of their primary star becoming a "real" star. The blast of stellar wind that blew most of the gas and dust particles from the vicinity of the new star forced much of it into the smaller swirls and eddies that were orbiting the primary eddy. Based on the large number of binary and triple star systems astronomers see in the sky, it is not unlikely that most solitary stars would have planets.

In fact, a study of the Orion Nebula by the Hubble Space Telescope detected protoplanetary disks around half of the 50 stars located in it. These disks, primarily dust and gas, appear to have a total mass of at least 15 times that of Jupiter.

Planets too close to the surface of the new star are scorched. Whatever water forms is quickly vaporized and driven into space. Similarly, any atmosphere is heated to such a high temperature that it boils away into space. Distant planets receive so little heat that water freezes solid as soon as it forms. Any atmospheres they may have are gases like methane, ethane, and nitrogen, with little or no oxygen.

Previously, accepted theory was that gas giants formed only in orbits where water cannot exist as a liquid on the surface, sometimes called the "snowline." Instead, the gas giants had hydrogen and helium gas atmospheres, with oxygen, carbon dioxide, and other gases as trace elements. If these planets were moved closer to their primary, the atmospheres would be heated to the point where the hydrogen and helium would simply boil away into space, leaving the heavier gases and any water molecules that may have formed.

However, the majority of planets discovered are gas giants in orbits well below this limit, with a couple barely outside the primary star's atmosphere (chromosphere) by a few million miles. It is now thought that gas giants form outside this limit in the dust cloud. Then, as their gravity sucks more mass from the cloud and interacts with the mass of the dust cloud, the gas giants slowly spiral in toward their primary. Some even drop into the star, increasing the amount of metals in the atmosphere (astronomers consider any element heavier than helium to be a metal). This could happen several times until enough material has been removed from the dust cloud to halt the process. Most of the stars with a discovered planet have a higher than normal "metallicity," but this seems to be more of a function of the higher metallicity of the dust cloud in which the star and planets formed.

This theory does present problems for terrestrial planets, however, as the giant planet should sweep clean the inner orbital area as it follows this death spiral, leaving little material behind for other planets to form.

Another interesting result of the planet discoveries is that many of the giant planets are in eccentric orbits, unlike the orbits of the planets in our solar system. Planet formation theory suggests that the orbits of the planets should be very close to circular, reflecting the natural state of the dust cloud orbiting the protostar. Eccentricity can be injected into the orbits by the interaction of the various planets once they reach critical sizes that can exert influence over other planets orbits, but this influence is usually restricted to the affect the largest planet has on the smaller ones. How did the giant planet get the eccentric orbit?

A possible explanation is that some of the giant planets formed concurrently with the protostar before the dust cloud had a stable orbital relationship with the protostar. Once the protostar made the transition to burning hydrogen it set the stage for normal planet formation according to traditional theories, but left the giant planets already formed intact.

This would explain the high eccentricities of many of the exoplanets. Support for this theory might come in the discovery of young (less than a few million years) substellar objects (less than ten times the mass of Jupiter) that formed in molecular clouds without the presence of a star. One such possible planet is S Orionis 70, in the belt of Orion. However, there is a counterargument that this planet is a two billion-year-old, 30 to 60 times the mass of Jupiter, massive, brown dwarf between Earth and the Sigma Orionis cluster. This last is strongly supported by the evidence.

The formation of atmospheres on terrestrial (rocky, as opposed to gaseous) cores is mostly conjecture, but many astronomers believe it is the result of the internal heating of the planet forcing gases to the surface through volcanoes. It may take up to 500 million years for an atmosphere to become established, assuming the planet has enough mass to keep the gases from boiling away into space. It might take only a few hundred thousand.

What happens next in the star's evolution is based on simple physics and observation.

The larger a star is, the more hydrogen it must burn to helium every second to keep from collapsing. This means that big stars, even though they have much greater supplies of hydrogen than smaller stars, go through their supply much faster. A star with a mass 25 times that of the sun will take only 7.5 million years to deplete its supply of hydrogen. A star with 15 solar masses will have a 15-million year supply. Two solar masses will yield a 500-million year supply, while a mass equal to the sun has a ten-billion year supply. Three-quarters a solar mass will give a 15-billion year supply, and half a solar mass will have a 64 billion year supply.

Given that current estimates are that it took 500 million years for a stable oxygen atmosphere to develop on Earth, a billion years for life to appear, and another three billion years for us to develop, finding inhabited planets around large, hot stars is not very likely. This does not mean that inhabitable planets will not be in orbit, just that the life forms are likely to be primitive (single-celled and small multicelled organisms), if they exist.

Very Old Stars Will Not Have Inhabited Planets

Stars have a limited supply of hydrogen. As the core fills up with helium, collisions between hydrogen atoms become rarer. Eventually, the hydrogen burning in the core reaches the point where not enough fusions take place to prevent a gravitational collapse. The star collapses until either the helium gets hot and dense enough to fuse, or enough hydrogen accumulates on the core's helium surface at a high enough pressure to resume hydrogen burning in a shell around the core. This repeats until the core gets dense and hot enough to fuse helium. Helium fusing releases even more heat than hydrogen fusing, so the star's outer region reverses its collapse and expands to many times its original size. The core, however, remains just as dense and hot.

When the sun depletes its hydrogen core and begins burning helium, it will expand all the way to Mars' orbit, vaporizing Earth. Jupiter's moons might become habitable by humans, and most of the atmospheres of

Jupiter, Saturn, Uranus, and Neptune will probably boil away. From a distance, the sun will appear to become much brighter. At this point the sun is a giant, and there is a tremendous outflowing of particles from its huge surface. The sun could lose up to half its mass during this stage of development.

This fuel-burning alteration and size change do not take place instantaneously. The exact time scale is not known, but stars such as Eta Carinae (an F5 star about 3,700 light-years away) have changed their size and brightness measurably over the last few centuries. Eta Carinae was the second brightest star in the sky in the 1830s and 1840s. Today, it is a ninth-magnitude star, less than one-millionth as bright as it used to be.

Virtually all giant and supergiant stars have passed through the hydrogen burning stage and are burning helium. Thus, any formerly habitable or inhabited planets around these stars will have been destroyed. The bigger gas giants, if they are outside the expanded star's atmosphere, should weather the expansion without difficulty. Their gravity is sufficient to keep their atmospheres from evaporating. The energy from the star cannot accelerate the individual molecules enough to exceed the escape velocity of the giant planet's gravity.

It is much more likely that a formerly uninhabitable frozen planet of Earth-like construction would thaw to the point of habitability. If a smaller giant planet does, indeed, "evaporate" under the heat of the newly expanded star, the loss of mass might be enough to allow the moons to escape and orbit the star as independent planets. Such a situation could lead to unusual situations of planets "sharing" an orbit. The distance between their orbital paths would be less than several times the diameter of the planets/moons themselves. There is no reason such planets could not exist around this class of star.

Unfortunately, Triton (one of Neptune's moons), Titan (one of Saturn's), and Ganymede, Callisto, Io, and Europa (of Jupiter's many moons), all appear to have rocky cores surrounded by layers of high and low density ice (water ice, methane ice). Those with atmospheres, Titan and Triton, use mainly methane and ethane as their surface liquids (*Astronomy*, April 1993, pgs 26–35). Heating up these planets would drive away most of the methane and ethane, and leave surfaces buried under liquid water with thick nitrogen atmospheres. Whether oxygen would appear in quantities sufficient for our kind of life is not known.

The time scale for helium burning is not nearly as long as that for hydrogen, and again it depends on the mass of the star. Massive stars will go through their supply very fast. Not-so-massive stars will last much longer. The time for stars the size of our sun appears to be only millions of years, not billions of years, rendering moot the possibility of life developing from single cell to intelligent.

Where Would the Planets Orbit?

How far from the star should planets orbit? At what points (too hot, too cold) does it become impossible for Earth-type life to exist?

At one time, astronomers thought there was a physical law that related the positions of the planetary orbits. Bode's Law* spurred the search for planets and resulted in the discovery of Uranus and the asteroids. Scientists now know that Bode's Law was one example of simple orbital physics and geometric progression. Mathematically, at least, if not observationally, this model works, but astronomers only have our solar system as an example. Scientists do not have enough information about the planetary systems of the newly discovered planets to draw conclusions, although several of the multiplanet systems seem to have orbital "resonance." That is, the orbital period ratios are a simple fraction (1:2, 2:3, etc.).

Bode's Law states that the orbiting planets in a system will interfere with each other, with the larger planets exerting the most influence. The orbital positions that become established reflect those influences. Planets too close together interact strongly and pull each other into new orbits. Eventually, orbits are established that offer the least amount of interference and appear relatively stable. The asteroid ring between Jupiter and Mars appears to be the result of Jupiter's massive gravity preventing a planet from forming and capturing for itself much of the matter that would have gone into this "missing" planet. If Jupiter had been smaller, there might be a planet instead of asteroids in that orbital position.

One astronomer made a random model of a star system and ran it with many starting orbits. He found that the systems usually stabilized into a situation where something like Bode's Law could be derived from the results. Nothing in the extrasolar planets we have so far discovered contradicts this.

The moons of Jupiter, Saturn, Uranus and Neptune seem to follow this pattern. Unstable orbits (over the course of millions of years) result in the problematic moon either crashing into another moon, spiraling into the primary it orbits, or being thrown clear of the orbital system entirely. Stable moons can be pushed into unstable orbits by the impacts of large asteroids or the nearby passage of another planet.

---

*The geometrical progression is based on powers of two. The first planet is R, the second is R+B, the third is R+(B*2), the fourth is R+(B*4), the fifth is R+(B*8), the sixth is R+(B*16), the seventh is R+(B*32), the eighth is R+(B*64), and so forth, with the number multiplying B doubling its previous value.

One extremely unusual orbital pattern was predicted in the 1800s by George Darwin (son of Charles Darwin) and was verified in the 1980s by close observations of Janus and Epimetheus, two then-newly discovered moons orbiting Saturn. These two moons co-orbit Saturn with an average separation of only 50 kilometers (at a distance of 151,000 kilometers from Saturn and time of 18 hours). This makes their orbital periods very close, with the inner moon having a period only slightly shorter than the other. (The closer a moon or planet is to its primary, the faster it moves in its orbit in comparison to objects farther out.)

Conventional thought is that eventually the two should collide or throw each other far enough apart that interaction is at a minimum. But that is not what happens. Instead, the two moons swap positions. The moon in the outer orbit moves into the inner orbit, and vice versa. This happens every time the moon in the inner orbit (which is moving faster) tries to pass the moon in the outer orbit. As the inner moon approaches the outer, its gravity drags the outer moon back, slowing it down. Simultaneously, the outer moon pulls on the inner moon, speeding it up. As the inner moon speeds up, its orbit moves outward. As the outer moon slows down, its orbit moves inward. From a distance, it looks as if gravity were actually pushing the two moons apart. This orbital swap takes place over a period of several orbits and once every four years (*Sky & Telescope*, April 1993, page 32).

For our needs, we can assume that Bode's Law is a general rule and adopt a formula based on it to help us place planets at realistic distances. The relationship of the orbits is $R_x = R + (B \times 2^{x-2})$ where R_x is the orbit of planet x in the system, R is the orbit of the first planet in the system, and B is the Bode Constant multiplied by 2 raised to the power of x minus 2. The Bode Constant is a number plugged into the formula (using a geometrical progression) to help establish the positions of the other planets. The Bode constant can be any number between .2AU and 1AU (AU stands for Astronomical Unit and equals 93 million miles, the distance from earth to the sun).

For our solar system, the Bode Constant is .3, and R, the first planet (Mercury), is .4AU, so the formula would yield a theoretical position for the third planet in the system of

$$\begin{aligned} R_3 \text{ (Earth)} &= .4 + (.3 \times 2^{3-2}) \\ &= .4 + (.3 \times 2) \\ &= .4 + .6 \\ &= 1AU. \end{aligned}$$

For the fourth planet in our system, the orbital position is

$$R_4 \text{ (Mars)} = .4 + (.3 \times 2^{4-2})$$
$$= .4 + (.3 \times 2^2)$$
$$= .4 + (.3 \times 4)$$
$$= .4 + 1.2$$
$$= 1.6 \text{AU}.$$

This is off by only .07AU (Mars is really 1.53AU from the sun).

If the star is a giant or supergiant, it is permissible for that first planet to be inside the surface of the star (such as in the case of our own system, when the sun begins burning helium and dramatically expands, a case in which the first three or four planets are "below" the surface). There will not be a planet there — the temperatures are too high — but for the math to work, the assumption is made that a planet is still there.

To create a theoretical star system, pick the numbers for a star system, and then compare the orbits with the diagram of the star's habitable zone. If none of the planets fall within the habitable zone for that star, choose a new first orbit, a new Bode Constant, or both, and recalculate. The starting numbers may be varied, but stick to the formula for the placement of the rest of the planets.

For example, HD 143761, Rho Coronae Borealis, has a giant planet. With the giant planet at .224AU and a Bode Constant of .5, the formula gives us planets at .224AU, .724AU, 1.224AU, 2.224AU, 4.224AU, 8.224AU, and 16.224AU in the first seven positions. Looking at the habitable zone diagram for the star, the second and third planets are both inside the zone. Planet two is slightly farther out than Venus, with planet three significantly farther out. This leads to the possibility of two habitable planets in one system.

The Greenhouse Effect

The greenhouse effect is based on observations astronomers have made on Earth and other planets that a planet with an atmosphere is warmer than a planet with no atmosphere. The atmosphere acts like a blanket, trapping part of the heat it receives from its sun.

Exactly how the gases in an atmosphere work to produce this effect and the contribution of each gas are not exactly known. From a practical point of view, the Greenhouse Effect theory states that increasing the water vapor or carbon dioxide content of the atmosphere of an Earth-like planet will cause the planet in question to be hotter than normal. Arguments abound that the Greenhouse Effect, combined with mankind's carbon dioxide production, is going to destroy Earth for people by melting the polar caps and turning the planet into a hothouse.

There are four major theories about what will happen if the amount of carbon dioxide increases dramatically.

One states that the increase will trap more of the sun's heat and raise global temperature by an average of one degree Fahrenheit. This, in turn, will melt the polar caps, flooding the coastal regions and greatly enlarging the desert regions. This is the theory that gets all the media attention.

The second theory also assumes that the global temperature will begin to increase. But instead of melting the polar caps, the increase in temperature will mean an increase in the water that evaporates from the oceans, lakes, and rivers. This additional water vapor will result in more clouds and storms, meaning that more of the Earth's surface will be hidden under clouds. Because clouds, being white on top, reflect more light and heat than land, plants, or oceans, there will actually be *less* light and heat reaching the ground, so the temperatures will begin to fall. Before it begins to fall, however, there will be *more* snowfall in the polar regions, increasing the sizes of the polar caps. As the clouds disappear, the ice will be revealed for a net zero change from the heavy cloud cover. The end result, instead of being a hothouse, will be a new ice age.

The third theory agrees in part with the second theory, but deviates at the increase in the polar caps. Instead, this theory says that the additional rainfall will result in the drastic reduction of the size of the world's deserts, replacing the lost coastal regions with now-fertile farmland. This increase in plant cover will reduce the reflectivity of the former deserts and balance the increased reflectivity of the clouds, striking a new equilibrium. Coastal regions would be lost, but the deserts would be habitable.

The final theory claims that the others have left out nighttime temperatures. According to this theory, the increase in carbon dioxide does not increase the daytime temperatures, the time during which the sun is pumping energy into the Earth. Instead, the carbon-dioxide slows the Earth from reradiating that accumulated heat into space at night. Thus, nighttime temperatures go up by one degree, but the average daytime temperatures do not. The net result is nothing changes except that the nighttime is warmer. There are no melting polar caps and no great increase in rainfall. Everything remains the same as it is today.

The important thing to remember about these theories is that they *all* use the same data to prove their points.

It will be at least 100 years before we can prove any of the popular Greenhouse Effect theories.

Binary Stars: Can There Be Planets?

Binary stars, two or more stars orbiting in the same system, do not act like planets orbiting the sun. By their very nature stars are huge. The planets, compared to the sun, make up less than one-thousandth of the total mass in the solar system, and most of that is Jupiter. The sun overwhelms the planets with its mass.

In a binary system, however, the two stars have a mass difference of usually less than 50 percent. That means the two stars orbit around a focal point, a common point on a line drawn through the center of the orbital path. If the two stars have exactly the same mass, then that point is centered between the stars. The effect is the same as tying two balls together with a string and then tossing them spinning in the air.

Orbits are rarely that simple. They can range from that simple case to one in which one star comes very close and then goes very far away, like Halley's Comet does to the sun, with an orbital period measured in the hundreds of thousands of years. Eccentricity is the mathematical term describing how far a circle deviates from true round, so astronomers use the same word to describe how far an orbit deviates from round. When the eccentricity equals zero, the orbit is perfectly round. The closer the eccentricity gets to one, the more elliptical the circle becomes. At one and above, the eccentricity describes a parabola. The arc in space that the Galileo spacecraft formed as it coasted by Jupiter on its way to Saturn was a parabolic orbit.

It was thought that any star system in which two stars approached each other closer than 30AU would not have stable orbits for planets. Recent computer models disprove this. The models indicate that stable orbits are possible in the region below one-fifth the distance of the two stars' closest approach.

So, if we have two sunlike stars no closer than 10 astronomical units, then stable orbits can exist, for each star, below 2 astronomical units. Both stars could have habitable planets in stable orbits. They would just have to be inside that two 2AU limit (for a system like ours, stable planets would include Mercury, Venus, Earth, and Mars).

Length of a Year

Determining the length of a planet's year is simply a matter knowing the mass of the planet's primary and the radius of the planet's orbit. If you use the sun's mass and the radius of the Earth's orbit as your standards, then the formula $P = $ Square Root of (D^3/M) where "D" is the orbital radius of the planet (cubed, or multiplied times itself three times) and "M" is the mass of the star it orbits, gives the time it takes the planet to complete one orbit ("P"), in Earth years.

To discover how long it would take a second sample planet to complete one orbit around HD 143761 plug in HD 143761's mass of 1.0 and the planet's .724AU from that star.

$$P = \text{Square Root of } (.742^3/1.0)$$
$$P = \text{Square Root of } (.408/1.0)$$
$$P = \text{Square Root of } (.408)$$
$$P = .638.$$

Multiplying .638 by the length of our year, 365 days, gives a rough figure of 232.87 Earth days for it to complete one orbit. This assumes that the planet is in a nearly circular orbit and not in an eccentric one that varies between .6AU and .7AU. Naturally, the actual length of a "day" on a hypothetical planet can be any number of hours, and the year can be any number of days. The figure given by the formula is just the length of the planet's year in Earth days.

Because only masses of double stars are known with any accuracy, assumptions about solitary stars must be based on their color type and size classification. There is a reasonable amount of error in these assumptions. These numbers are guidelines only. A planet may be heavier or lighter because of mass measurements inaccuracies.

More Evolution

Big stars evolve quickly. As they evolve, they change size and brightness. Our sun has increased its brightness by 30 percent in the last four billion years. This gives us a pitfall in planet-building. Using our system as an example, three planets are in the life zone — Venus, Earth, and Mars — but only one is habitable.

When the sun's supply of core hydrogen runs out, it will collapse until helium burning takes place, leaving carbon and oxygen in the core. It will expand until it is as large in diameter as Mars' orbit. It will drive off a measurable amount of its mass in a great solar wind.

The former habitable planets will be vaporized, but the gas giant planets — Jupiter, Saturn, and so on — will warm up. The moons, especially Titan (orbiting Saturn), will undoubtedly be available for humans. (Of course, Titan's methane, ethane, nitrogen atmosphere must be converted to the correct oxygen/nitrogen mix.)

Gradually, the sun will run out of helium in the core and begin collapsing again until the hydrogen compresses enough to reignite in a shell around the core, replenishing the helium. At some point, the helium fires up again and stops or slows the hydrogen fusing. The star may oscillate for thousands, or hundreds of thousands, of years. Each time it switches from hydrogen to helium fusing, it blows more of its mass away, forming a shell of expanding gases called a planetary nebula (actually a globe of expanding gases that looks like a ring when seen through a telescope).

This oscillation can have a period from hours to months to years, depending on the size and mass of the star involved. The change in brightness may be a thousand times different. Such stars are called variables.

Eventually, the star runs out of hydrogen and helium in the core area, and the star begins a final collapse.

At this time the star's mass becomes important. Stars more than three times more massive than our sun, at this point, move to the next stage and start fusing the carbon and oxygen in the core to make iron. This results in a supernova. The star loses up to 90 percent of its mass in a stupendous explosion. The remnant is very dense (less than 15 miles in diameter) and leads to such peculiarities as pulsars and neutron stars. Stars not massive enough to burn carbon efficiently for a supernova will collapse into black holes.

Stars of three solar masses or less can end up in many forms, all a category of the subdwarf, or degenerate, stars. They are called white dwarfs.

White dwarf stars, while very hot, probably will not have any planets closer than 5AU. They would have been destroyed millennia ago. So, we have the so-called white dwarf, a star radiating the leftover heat of its collapse and glowing so hot it is white in color. A closely orbiting planet could easily be habitable by humans, but no such planets could exist. Even if a planet could survive the nova, there is the problem of orbital dynamics. Typically, the star loses over half its mass in the nova. When a star's mass is decreased, the planets orbiting it drift into orbits much farther from the remaining core. The planets farthest from the star might leave the system completely. A planet would have had to exist inside the prenova star to have a habitable planetary orbit around the white dwarf. However, planets orbiting pulsars have been found.

There is no such thing as a "dead" star. Mathematical estimates show that it will take more than 100 billion years for even the smallest, dimmest red white dwarf star to cool off. Given that the age of the universe is only about 15 billion years, we have a long time to wait for the first white dwarf to cool down enough to land on it.

The same holds true for the brown dwarfs. They have such a small surface area and such high residual internal temperature that it will be billions of years before they cool down to the temperatures we find on Jupiter.

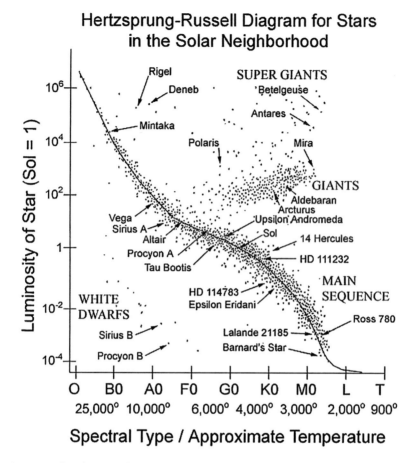

Figure 2. The above is the Hertzsprung-Russell Diagram, a visual guide to the spectral types based on observation of the stars we see.

Stellar Classes

Astronomers created an observation-based classification system using the stars' colors and sizes. The colors are ranked as O, B, A, F, G, K, M, L, T (an easy way to remember the order is: Oh Be A Fine Girl, Kiss Me Lovely Thing). The O (blue-white), B (almost white), A (white), and F (yellow-white) classifications are the bright white stars, G-types are yellow stars, K-types are orange, and M-types are red. L and T-types are the new designations for brown dwarfs. Each class is subdivided with the use of a number following the letter.

Originally, the plan was to have ten subdivisions in alphabetical order. Errors in assigning colors, and chemical differences that were discovered

Classes of Stars by Spectral Type

Type	Color	Surface Temperature	Main Characteristics
O	Blue	> 25,000 K	Strong ultraviolet continuum. (10 Lacertra Mintaka, aka delta Orionis)
B	Blue-White	11,000 - 25,000	Neutral helium lines in absorption. (Rigel, Spica)
A	White	7,500 - 11,000	Hydrogen lines at maximum strength. (Sirius, Vega)
F	White to Yellow	6,000 - 7,500	Metallic lines become noticeable. (Canopus, Procyon A)
G	Yellow	5,000 - 6,000	Solar-type spectra. (Sun, Capella)
K	Orange	3,500 - 5,000	Metallic lines dominate. Weak ultraviolet. (Arcturus, Aldebaran)
M	Red	2,000 - 3,500	Molecular bands of titanium oxide. (Betelgeuse, Antares)
L	brown	1,500 - 2,000	Some to little Methane
T	brown	< 1,500	Methane detected easily (Gliese 229B)

Figure 3. Stars are divided into types based on their colors.

to be superficial, doomed the original scheme. For example, the B stars are broken into B0, B0.5, 1, 2, 3, 5, 8, and 9. The G-type stars are subdivided into G0, 2, 5, and 8. Uncertainty is shown by listing the letter without a modifying number. Earth's Sun is a G2-type star. Subdivisions are added as stars are found that fit them.

The spectral type is a direct function of the mass of the star. The more massive a star is, the hotter and bigger it is. Ironically, the more massive a star, the shorter its lifespan. This is because much higher internal temperatures are required to prevent its collapse, thus it burns through its internal fuel supply very rapidly. For example, the type "O" stars typically have lifespans measured in tens of millions of years while the "K" and "M" type stars have life-spans in the tens of billions of years.

Color alone, astronomers found, was not enough, for it is possible for a star to be a supergiant O5 or a dwarf O5. Both have the same color but are dramatically different in size and age. Astronomers append the Roman numerals 0 (extreme supergiants), Ia (luminous supergiant), Ib (less luminous supergiant, II (bright giant), III (normal giant), IV (Subgiants), V (mainsequence, or dwarf), VI (subdwarf, also known as sd), and VII (white dwarfs, also known by the designation "D" or "wd"). Dim or unusual dwarf stars are usually prefixed with a "d" and the size letter "V" is left off.

The designations VI and VII are less common and will probably be dropped in favor of the alternates "sd," "D," and "wd." Our sun is a G2 dwarf, as is Alpha Centauri, our closest neighbor.

The blank spots in the number system reflect one of two possibilities: either it is flatly impossible for a star of that color and size to exist, or that it is such a short stage in stellar evolution that stars do not stay in that stage long enough for us to have observed them yet.

The "sd" dwarfs are known to belong to the metal-deficient Population II stars. A Population II star is a member of the great halo of dust and stars that encircle the Milky Way in a vast globe. A Population II star is very old and part of the first generation of stars to form in our galaxy. As a result, it has little or none of the heavier elements such as lithium, carbon, oxygen, silicon, and other metals. Any planets formed around it would be hydrogen and helium mixes only. Extremely few atoms heavier than hydrogen existed where and when this star formed. No habitable planets of any kind can exist around a Population II star. In fact, no life forms of any kind could have evolved around these stars. The complex organic (carbon, oxygen) molecule chains require atoms higher on the atomic chart than hydrogen or helium, and their lack around these stars is fundamental.

The "wd" stars run the range of the spectral sequence, from O (white) to M (red). However, because the first one discovered was a white dwarf, all of them are called this no matter their true color. These stars, about the size of the Earth, are all smaller than the lower size limit needed to ignite nuclear fusion, which is about the size of Jupiter, even though they have masses hundreds of times that of Jupiter. Therefore they are the remains of larger stars that have evolved down to this stage.

Classes of Stars by Luminosity		
Class	Description	Examples
Ia	Bright Supergiants	Rigel, Betelgeuse
Ib	Supergiants	Polaris (the North star)
II	Bright Giants	Mintaka (delta Orionis)
III	Giants	Arcturus, Capella
IV	Subgiant	Altair, Achenrar
V	Main sequence	Sun, Sirius A
d	White dwarfs	Sirius B, Procyon B

Figure 4. In each spectral classification, stars are subdivided by how much light they give off when compared to our sun. Rigel, a B-type blue, is 10,000 times brighter than our sun. Sirius B, an A-type blue, is 1/1000 as bright as our sun.

In some charts, suffixes are added to the classifications. These are technical details regarding special characteristics and are beyond the scope of this book. Refer to the texts in the appendix for further reading materials.

All stars in the classes 0, Ia, Ib, II, III, and IV are stars that have depleted their supply of core hydrogen and evolved off the main sequence. Any habitable planets they may have had in close orbits have been destroyed.

A Galactic Habitable Zone

Just as there is a narrow zone around a star outside of which no habitable planets exist, recent research has concluded that a similar zone exists on a galactic scale.[95] Stars located too close to the galactic center are subject to interference from other stars. The problem is not just that "close" encounters with other stars would seriously perturb planetary orbits and eject the planets from the system, but that the frequent near passage of other stars will perturb the Oort-cloud, that disk of debris orbiting at the outer edges of all stars. Containing trillions of comets, each near passage to the Oort cloud would send hundreds, if not thousands, of them falling into the planetary system, greatly increasing the chances that one of these would strike any habitable planets. With such impacts occurring every few million years, life would have difficulty surviving. Too many impacts, and the planet would develop such an eccentric orbit that it would begin to interact with the other planets in the system and eventually be ejected or thrown into its primary.

With the stars so closely packed together, the odds of a supernova event wiping out all traces of life are also much higher. After the supernova, the nova remnants (pulsars) would send out great beams of deadly X-rays capable of sterilizing planets hundreds of light-years away as they sweep over them. The center of the galaxy is awash in harmful gamma rays, X-rays, and cosmic rays, constant bombardment by large meteors, and close approaches by nearby stars.

The centers of the galaxy's spiral arms are just as deadly. The density of gases and interstellar matter allowing the creation of new stars makes them dangerous as radiation generated by protostars and large-scale gravitational fields would disrupt planetary life and orbits. While the stars are not as closely packed as the core area, there are still many pulsars making habitable planets rare.

Stars with habitable planets cannot have an eccentric galactic orbit or they stray too close to the galactic core or cut through the spiral arms too frequently. Our sun has a nearly circular orbit that keeps it a safe distance from the core and at the same time keeps us in pace with the spiral arms.

The older stars making up the majority of the globular clusters and core area of the galaxy are "metal-poor." They formed when the primary ingredients available were only hydrogen, helium, and some lithium. Not enough of the heavier metals like silicon, iron, or oxygen were available for rocky planet formation.

Lending support to the idea that closely packed stars are not conducive to planets was a survey of the globular cluster 47 Tucanae.[96] More than 34,000 stars were examined over eight days for indications of Jupiter-size planets crossing in front of their primary. Blind computer simulations with planets inserted at the rate seen in our galaxy proposed a recovery rate of 60 percent. It was expected that the survey would find 15 to 20 planets. None were found.

Similarly, the Greater and Lesser Megellanic Clouds, those two large irregularly shaped star clusters orbiting our Milky Way galaxy, are being examined for planets.[97] So far it appears that the chance for terrestrial planets is slim. It seems that, on average, stars in both clouds are Population II stars with very low metallicity, too low for terrestrial planetary formation. Second-generation Population I stars might form with planets, but they would be rare.

There appears to be a band starting at 4,500 parsecs (14,670 light-years) and going out to 11,500 parsecs (37,490 light-years) from the galactic center, with gaps where it crosses spiral arm structures, in the disk of the galaxy where habitable planets should be found. The "halo" stars above and below this disk are older, metal-poor stars.

Finally, there is galactic age. When the Milky Way first coalesced, large numbers of stars in the disk formed, burned through their fuel, exploded, and sprayed the galaxy with heavier elements. Our sun was a beneficiary of this earlier generation. The available material has been gathered up by other stars and there are actually fewer heavier elements available today than there were four billion years ago. Also, the radioactive material that provides the hot core of earth, and many of the heavier metals, is decaying. A planet forming today would have 21 percent less radium than Earth. A planet that formed substantially before Earth will have less iron and other heavy elements, and a planet that formed later than Earth will have the same problem.

It appears that Earth, at 4.5 billion years old, is at the tail end of optimal planet formation. The "window" for planet formation, where the planet has reasonable amounts of metals, began about five billion or six billion years ago. Any sooner, and all the deadly supernovas would have wiped out life. Any later, and we would not have the industrial capability we have developed.

Pulsar Planets

A pulsar is a supernova remnant with a mass up to twice that of the sun yet with a diameter of only 20 kilometers (14 miles). Because of their small size (compared to the pre-supernova stellar diameter of millions of miles), pulsars spin very fast, from thirty to hundreds of times per second. To the surprise of everyone, some pulsars appear to have planets.

Pulsars emit a radio pulse with every rotation. Actually, they are emitting a constant beam of X-rays from their magnetic poles. Because the magnetic pole is not aligned with the rotational pole, a "blip" of X-rays is seen as the pulsar's rotation swings its magnetic pole across our line of sight. If Earth were much closer to the pulsar, the X-ray blast would kill anything it hit. With a spin rate of 30 times a second it would not take long to sterilize a planet.

Tiny variations in the spin rate of two pulsars (PSR 1257+12[87] and PSR 1620–26[88]) have convinced two research teams that there are planets around them. PSR 1620–26 has a white dwarf companion with an estimated age of 500 million years.

Current theories suggest that PSR 1257+12's planets formed out of the remains of its supernova event (research has confirmed supernova events are not symmetrical). The asymmetry implies that sections of the stars' mantle may be thrown out in chunks large enough to coalesce into planets. Or perhaps a significant portion of the heavier metals formed in the event might be trapped into a disk orbiting the new pulsar and thus undergo conventional planet formation dynamics. These planets cannot be leftover from the prenova stage as they would have had to orbit inside the star's surface (also, remember, a star can lose more than half its mass during the red giant stage, and a planet's orbit would slowly drift farther away from the primary).

The PSR 1620–26 planet seems to be the result of a capture event. This pulsar lives in the old globular cluster M4, about 7,200 light-years away. The conjecture is that the planet formed 13 billion years ago orbiting a yellow star at about the same distance Uranus is from our sun. As the planet aged it survived the ravages of blistering ultraviolet radiation (from the other stars in the cluster's core), supernova radiation, and subsequent shockwaves in the furious firestorm of the young cluster. About the time multicelled life appeared on Earth the star and its planet were plunging into M4's core. In this relatively crowded space the star and planet passed close to an ancient pulsar, formed in a supernova when the cluster was younger, that had its own stellar companion. In a slow-motion gravitational maneuver the star and planet were captured by the pulsar while the original companion was flung from the nebulae.

Recoil from this action threw the new triple system into the outskirts of the nebulae. Eventually, as the yellow star aged, it became a red giant and spilled matter onto the pulsar, transferring momentum to the pulsar and speeding it up until it became a millisecond pulsar (rotating hundreds of time per second). The planet, meanwhile, continued on its leisurely orbit two billion miles from the two stars.

Gas Giant Planets, Inhabitable Moons?

The concept that gas giant planets could have moons with breathable atmospheres and liquid surface water, long the favorite of science fiction movies, was considered scientifically impossible. Theories of planet formation stated gas giants formed at the minimum distance at which hydrogen, helium, methane, and other light gas atoms were "blown" by the primary star's gas outflow and photon pressure (also called the "stellar wind"). This theoretical distance was well beyond the distance at which water is liquid on a terrestrial planet's surface. The discovery of "hot Jupiters" has disproved that theory. It is now acknowledged that giant planets either form beyond this boundary and then spiral closer, or form concurrently with their primary with an eccentricity that may be reflected by the dust cloud orbiting the developing star. Now the only theoretical constraint is moon size and orbit, and the orbit of the gas giant.

Computer simulations[90] have yielded several conclusions, assuming a planetary mass that is at least equal to Jupiter. First, there is no upper mass limit for moons of planets orbiting more than .6AU from their parent star. Second, stable moons can exist with these minimums:

Mass of Moon	Planet Orbit	Star Mass
Moon	.06AU (5.8 million miles)	$.24M_\odot$
Mars	.1AU (9.3 million miles)	$.30M_\odot$
Earth	.17AU (15 million miles)	$.38M_\odot$

Thus, for an Earth-sized moon orbiting a Jupiter-sized planet the minimum orbital distance to the primary star is .17AU. What is less obvious is that below the star mass minimum, the moon's orbit becomes unstable and it either spirals down into the gas giant planet or is lost to space or the star in less than five gigayears.

If the gas giant has a mass equal to Saturn, these minimums become:

Mass of Moon	Planet Orbit	Star Mass
Moon	.10AU (9.3 million miles)	$.3M_\odot$
Mars	.19AU (17 million miles)	$.4M_\odot$
Earth	.30AU (28 million miles)	$.52M_\odot$

These numbers are not straight lines on a graph, but curves. In general, as the mass of the planet increases, so does the maximum mass of the moon at any given orbital range and for any selected star mass. In all cases, any gas giant orbiting at .6AU or higher has no upper limit on the mass of any of its moons.

Lagrange Points

Lagrange points are locations in space where gravitational forces and the orbital motion of a body balance each other. They were discovered by French mathematician Louis Lagrange in 1772. He postulated that there were five such points. Three are on a straight line from the primary object to the orbiting object, the other two are in the orbit of the object, but 30 degrees in front and behind the object.

His discovery was astronomically confirmed in 1906 with the discovery of the Trojan asteroids orbiting at the Sun-Jupiter L_4 and L_5 points. More recently the Voyager probes found tiny moonlets at the Saturn-Dione L_4 point and at the Saturn-Tethys L_4 and L_5 points.

An object at L_1, L_2, or L_3 is almost stable, like a ball sitting on top of a hill. A little push or bump and it starts moving away. A spacecraft at one of these points has to use frequent, small rocket firings or other means to remain in the area, as the gravity of other planets exerts a pull over the course of an orbit. Orbits around these points are called "halo orbits." The Solar and Heliospheric Observatory (SOHO) is in a halo orbit around the Sun-Earth L_1 position, about a million miles sunward from Earth, and the Microwave Anisotropy Probe (MAP) is in a halo orbit around the Sun-Earth L_2 Position, about a million miles in the opposite direction.

An object at L_4 or L_5 is truly stable, like a ball in a bowl: When gently pushed away, it orbits the Lagrange point without drifting farther and farther away, and without the need of frequent rocket firings. The sun's pull makes any object in the Earth-Moon L_4 and L_5 locations move around, or "orbit," the Lagrange point in an 89-day cycle.

The Lagrange points are of interest in extrasolar planetary systems as they hold the possibility of harboring habitable terrestrial-sized planets in those systems where the gas giant planet is in the habitable zone of the primary star. Preliminary studies[89] seem to indicate that low eccentricity gas giants would allow terrestrial planets inside these lagrange points to survive with eccentricities between zero and .15. Eccentricities higher than .15 for the terrestrial planets ends up taking them outside the habitable zones.

Water Worlds

Recently, it has been proposed that there are planets between one and eight times the mass of the Earth that form outside the "snowline," that is, 5AU to 10AU where hydrogen, helium, and methane are pushed by the star's photon wind, and migrate inward to the habitable zone on a timescale of about one million years.[82] The planets are mostly made of refractory material (metals and silicates) and ices. From what we know of the composition of protostellar nebulae these should be about 50–50 metals/silicates and ices, by mass. Because of their small size and closeness to their primary, the planets will contain only trace amounts of hydrogen and helium. Most of the ices should be water.

A consequence of this scenario is that at high gas pressures (10 to 60 atmospheres, extrapolated to higher levels), carbon dioxide ice is heavier than water ice and will end up locked into a solid ice mantle covering the rocky planet core. Another is that water has ten ice phases as the pressure increases up to 128 atmospheres, each of increasing density, with only the first one less dense than liquid water.

The paper[82] specifically examines a six-Earth-mass planet, which has a diameter twice that of the Earth. It has a 100 kilometer thick ocean on top of a 4,739 kilometer thick mantle of ice (a combination of water ice and carbon dioxide ice, with the carbon dioxide at the bottom). The inner 7,895 kilometers are metals and silicates, with an iron core at a temperature of about 10,000 degrees K (compared to Earth's core temperature of about 5,000 degrees K). The surface gravity at the top of the ocean is 1.54 times that of the Earth. This assumes a 7 degrees C (45 degrees F) average surface temperature, the same as the Earth. A higher temperature would make for a deeper ocean, such as 133 kilometers at 30 degrees C (86 degrees F). The ocean bottom is the point at which the temperature is 0 degrees C. Interestingly, the gravity at that point is 1.96, so the deeper one goes in the ocean, the higher the gravity becomes. The pressure at the ocean/ice mantle interface is 250 times that at the surface.

The heat from the rocky mantle melts the inner ice interface upward at a rate of about one kilometer per million years, creating an inner ocean many times deeper than the surface ocean.

One possibility not mentioned in the paper is that there might be a narrow layer of metals and silicates deposited at the bottom of the ocean by meteor and comet impacts. Earth, for example, receives approximately 110 tons of metals and silicates from micro-meteor impacts every year. Multiply that over a billion years, add in some large asteroid impacts, and a substantial amount of dissolved metals will be formed in the water as well as a thin layer resting on the ice mantle.

Naming Conventions

Astronomers have divided the sky into 88 constellations. By definition, every star in the sky is in one constellation or another. Only 48 of the constellations were named by the ancient astrologers. The rest were all defined and named in the seventeenth and eighteenth centuries. Appendix C lists them all.

As a matter of convenience, astronomers decided to name stars in each constellation using the Greek alphabet, and assign the brightest star the name Alpha, the second brightest Beta, the third brightest Gamma, and so on. For example, a star named Beta Cygni means simply that it is the second brightest star in the constellation Cygnus, The Swan. While the name of the constellation is Latin (Cygnus), the genitive form of the name (Cygni) is used for the star's name.

Naturally, there are more than 24 stars in each constellation, so astronomers started using numbers (ignoring the first 24 that would be the numerical positions of the Greek letters). This would have worked nicely, except that individual astronomers, and several groups, started compiling lists of stars for special purposes. They frequently assigned previously named stars new designations.

For example, the Royal Greenwich Observatory created the *Catalog of Stars Within 25 Parsecs of the Sun* (1970). In it, the authors simply numbered the stars sequentially. R.G. Aitken compiled a list of multiple stars (stars close enough to orbit each other, and so form one system), and gave each group of orbiting stars one identifying number, the ADS series. R.A. Rossiter compiled a list of dim red stars, the Ross series. Professor Wolf released a list of unusual dim stars, the Wolf series. Plus, there are the Bonner Durchmusterung (BD), Cordoba Durchmusterung (CD), the Cape Photographic Durchmusterung (CPD), and Henry-Draper (HD) lists. In all, there are at least two dozen lists. Thus, it is not unusual for a star to be known by multiple names and numbers.

In some books, television shows, or in movies, the nomenclature of Algol III is used as a reference to the third planet of Algol. This is not the way astronomers name stars or planets. Astronomers have simply extended their naming conventions to include planets, with the difference that planets are designated with lowercase letters. That is, the primary star is labeled "A." If the system has a second star, it is labeled "B." If there is a planet it would be labeled with "b." Thus, the first planet around HD 209458 is listed as HD 209458b and the primary is referred to as HD 209458A. In a binary system, such as HD 114762, the first planet around HD 114762A would be referred to as HD 114762Ab. The first planet around HD 114762B, the companion star, would be referred to as 114762Bb.

Constellations and Perspective

One of the fallacies of pulp science fiction is that the constellations change dramatically as you move from star to star. This comes about from the belief that the closest stars to us make up the constellations. They do not. Almost one third of the stars that make up the major components of each constellation are located hundreds or thousands of light-years away. Andromeda galaxy, that faint patch of fuzzy light in the constellation of the same name, is more than two million light-years away.

The other fallacy is that the constellations do not change at all as the viewer moves about. This comes from the belief that the stars in the constellations are all so far away that they do not change positions when the observer moves from one star to another. This also is untrue. Sirius, for example, is the main star in the constellation Canis Major, the Larger Dog, of Orion, the hunter. It is only 8.65 light-years away. Even the small movement of going to Alpha Centauri would dramatically change Canis Major, as Sirius would appear to have moved a considerable distance across the night sky.

The reality is that constellations are made up of stars at many distances. It is somewhat like living in a city surrounded by mountains. Moving from one house to another across the street does little to change the overall view of the mountains, although it may significantly change the view of the neighborhood. But the farther the observer moves, the more the view of the mountains will change. And change is dependent entirely on the direction moved: Moving north or south changes the view of east and west much faster than it changes the view of the north and south.

The pulp hero/heroine will not look at the night sky and see nothing familiar. Unless the observer is on a planet very far from Earth, some constellations will change only slightly. Moving from Earth to Mars will not change the view of the night stars at all; the distance moved is too insignificant. Time is also a factor. The further the viewer is from our current time the more the night sky will change. For example, orange-red Arcturus is the most prominent star in the spring sky, in the constellation Boötes. This star is moving, obliquely, in our direction, at about 5,000 mph. It is now almost at its nearest point to us, but in a few thousand years, though still very bright, it will no longer be the brightest star in Boötes. In a million years, it will disappear from our sight. Other nearby stars will likewise move closer and then recede. Over the short term (a few thousand years), most stellar movements will be unnoticeable.

Definitions and Measurements

All measurements in this book are a best guess given the information available. All star mass measurements can vary by 10 percent depending on which planet search group did the calculations and assumptions it made. Masses of solitary stars are based on several things, including the star's apparent magnitude, its distance, the ratio of iron-to-hydrogen (Fe/H — metallicity) in its atmosphere, and its spectral type. The mass of the star directly affects the measurements of the planets orbiting it. If the star is bigger than astronomers think, so is the planet; if smaller, the planet is smaller. The ages given for the stars, their diameters, radii, and temperatures are just as ambiguous. The most often cited source for distance measurements is the Hipparcos project, a satellite that examined more than 118,000 stars for several years. Unfortunately, a booster failure left the satellite in an elliptical orbit and caused data accuracy problems. A mathematical algorithm was applied to the data to try to correct for the errors, but opinions differ as to the success of this operation. We do not know the distances as precisely as we like to pretend. It is not unusual for calculated distances to vary plus or minus five light-years or more.

Myr   Megayear, a million years, 1,000,000, 1×10^6
Gyr   Gigayear, a billion years, 1,000,000,000, 1×10^9
$M_\odot$ Mass of our sun (1.989×10^{30} kilograms)
$L_\odot$ Luminosity of our sun, 3.90×10^{26} watts (Earth receives 1,340 watts per square meter when the sun is directly overhead; for example, at the equator on March 21 and October 21)
$R_\odot$ Radius of our sun = 695,000 kilometers, 434,000 miles
$D_\odot$ Diameter of our sun = 1,390,000 kilometers, 868,000 miles
M_J Jupiter's mass = $.000955 M_\odot$, 1.899×10^{27} kilograms
D_J Jupiter's diameter = 142,984 kilometers, 88,789 miles
M_E Earth's mass = $.000003 M_\odot$ = $.003141 M_J$, = $.0105 M_S$
M_S Saturn's mass = $.000285 M_\odot$, $.298 M_J$
M_N Neptune's mass = $.0005128 M_\odot$ = $.0536 M_J$ = $.184 M_S$
AU   Astronomical Unit = 149.6 million kilometers or 92,960,116 million miles
Lyr   Light-year = 63,240AU = 5,878,797,735,840 miles
K   Kelvin, the absolute temperature scale. 273.15°K = 0° Centigrade, 32° Fahrenheit (water freezes); 373.15°K = 100°C, 212° Fahrenheit (water boils)
ecc.   Eccentricity, a measurement how circular an orbit is, with zero being a circle.

semi-major axis   The radius measurement of a circle along the line of its widest section:

FeH   Metallicity, the ratio of iron atoms to hydrogen atoms in the surface layer/atmosphere (chromosphere) of a star.

R.A.   Right Ascension. Just like the Earth is divided into 24 time zones (24 hours per day), astronomers divide the sky into 24 slices of one hour each. They measure the positions of the stars by how far west of Greenwich, England, they are at midnight on the night of the vernal equinox (March 21). Right Ascension (R.A.) is a measure of how long one would have to wait for a particular star to reach its zenith over Greenwich, England. Thus, a star with a Right Ascension of 10 hours, 15 minutes, 6 seconds means that 10 hours, 15 minutes, and 6 seconds after midnight on the night of the vernal equinox, that star reaches its zenith in the sky over Greenwich, England.

D   Declination, the measure of how far above or below the Earth's equator a star is located. Positive numbers indicate the star is in the northern hemisphere's sky, negative numbers indicate the southern hemisphere.

Apparent Magnitude   How bright an object appears in the moonless night sky. A 1st magnitude star is roughly 100 times brighter than a 6th magnitude star and thus each magnitude corresponds to an increase of 2.51 times in the star's brightness. This means that a 1st magnitude star is 2.51 times brighter than a 2nd magnitude star. The faintest objects that can be seen with the naked eye are of magnitude 6.5 Some typical magnitudes are:

> Sun -26.5
> Full moon -12.5
> First quarter moon -10.20
> Last quarter moon -10.05
> Venus -4.6
> Jupiter, Mars -2.9
> Sirius -1.5
> Naked eye limit 6.5
> Binocular limit 10
> Hubble Space Telescope limit 29

Planetary Discoveries

How to read the planetary charts.

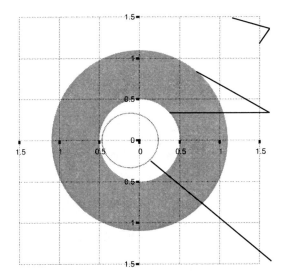

The tic marks are the distances from the primary in Astronomical Units.

The shaded area is the calculated habitable zone for the star (the area in which water could exist as a liquid on the surface of the Earth-like planet).

The calculated orbit for the planetary candidate.

   The habitable zone calculated for each star is based on estimates of its light output, which depends on the size, type, distance, and age of the star, all of which are estimates. The orbit calculated for the planetary candidate is dependent not only on the accuracy of the orbital period measurements made by the discovery team, but on the estimated mass of the star and the inclination of the planet's orbit to our line of sight. Some of the candidates presented are probably more massive brown dwarf stars in highly inclined orbits. However, the information presented on all the planetary candidates and their primaries represents the best possible data at this time.

HD 114762

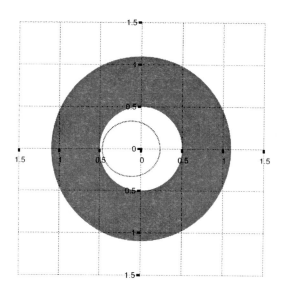

HD 114762 is located in the constellation of Coma Berenices, Berenices Hair, and is a type F9V star, an old halo star. It has a measured distance of 132 light-years from Earth (RA 13 12 19.74, D +17 31 01.64), a 7.3 apparent magnitude, a $.80M_\odot$ mass, a 5,884°K effective temperature, a $-.70\pm.04$ FeH ratio, and a $1.69L_\odot$ luminosity, indicating it has evolved off the main sequence and is a giant star.

HD 114762 may be oriented nearly pole-on to Earth, making estimates for the planet difficult to establish. Given that it is an old halo star, and has spun down (its rotational speed is much less than our Sun), the uncertainty could be large. As an old halo star it has a low metallicity and lower mass than expected (from what the apparent magnitude suggests, it has a larger than normal diameter).

As a result, the planet, the first reliably detected (1989, Geneve Observatorire[1]), could be anywhere from an $80M_J$ brown dwarf to a $9M_J$ planet. Present measurements, based on a $9M_J$, give an $84\pm.134$ day period, a .351AU semi-major axis, and a $.33\pm.01$ eccentricity. Plausibly, based on the low metallicity and old halo status of the primary, a constraint of 9_J-$11.5M_J$ is placed on the planet's size. Most astronomers mark it down as a brown dwarf.

Unfortunately, the old-halo status of the primary reduces the probability of any terrestrial planets in the system. Virtually all the planets in this system would be gas giants. The system should be empty of the rocky debris we find in our solar system. Any other planets would have to orbit above .7AU to escape being ejected from the system. Orbital simulations show that out of 100 test planets, only one is still in orbit after one million years.[79] Underwood, et al., report that a terrestrial planet orbit could remain in the outer portion of the habitable zone for the last one billion years.[86]

HD 217014 (51 Pegasi)

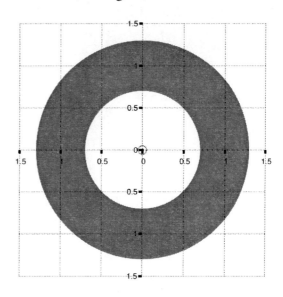

HD 217014 (51 Pegasi) is the 51st brightest star in the constellation Pegasus. It is a type G2IV star at a distance of 47.8 light-years (RA 22 57 27.14, D +20 46 04.5) with a 5.49 apparent magnitude, a $1.05M_\odot$ mass, a $5,805\pm36°K$ effective temperature, and is slightly evolved off the main-sequence. From its metallicity ($.20\pm.05$ FeH ratio) an age of 6–8 Gyrs is estimated.

It is a solar analogue; that is, its spectrum is almost identical to our Sun, except for the much higher metallicity (if you put a chart of its spectrum over one of our Sun, they would be indistinguishable).

51 Peg and the Sun are quite similar in size, luminosity, and mass. Its rotation is between 29.6 and 37 days with a radius between 1.1 and $1.3R_\odot$. It might be a variable star; i.e., its light output oscillates with time.

The planet, the second discovered (1996, Lick Observatory,[2] Keck Observatory[2]), is very close to the primary with a $.052\pm.001$AU semi-major axis (less than 4.5 million miles from its primary), a 4.231 day period, a $.01\pm.003$ eccentricity, a $.46\pm.02M_J$ minimum mass, and a 1,250°K temperature. If it were a solid planet instead of a gas giant its radius would be about $.3R_J$ or about 13,404 miles.

Measurements indicate that there are no other large planets orbiting the star. Given the close orbit of the giant planet, a terrestrial planet could be placed anywhere inside the habitable zone.

Orbital simulations show that out of 100 test planets, eighty are still in orbit after one million years (simulations of our solar system yields only eighty-one out of 100).[79] Underwood, et al., report that a terrestrial planet orbit could remain in the habitable zone for the last one billion years.[86]

HD 117176 (70 Virginis)

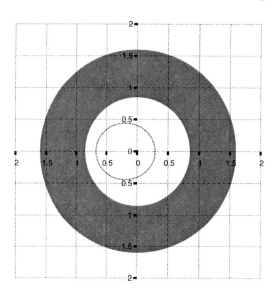

HD 117176 (70 Virginis) is located in the constellation of Virgo, The Virgin, and is the 70th brightest star in that constellation. It is a type G4V star 59 light-years from earth (RA 13 28 26.54, D +13 47 12.43), with a 5.0 apparent magnitude. From its metallicity (-.06±.05 FeH) an age of 8 Gyrs is estimated.

70 Vir. and the Sun are similar. 70 Vir. has slightly lower effective temperature (5,560°K), $2.86L_\odot$ luminosity, and is a main-sequence star with a $.92\pm.18M_\odot$ mass. However, the apparent magnitude suggests that 70 Vir. has begun to evolve toward subgiant status, with an implied age of 8 Gyr. Consistent with this old age, it is chromospherically inactive (no sunspots or flares) and rotates with a 35±7 day period, slower than our Sun's 25.38 days.

The planet, the third discovered (1996, Lick Observatory[3]), has a .482±.05AU semi-major axis, a 116.7 day period and a .40±.01 eccentricity. Moons up to the size of Earth would have stable orbits, but the radiation belts and close orbit would render them uninhabitable. Its surface temperature is 360°K (water boils at 373°K). Models that include internal and external heat sources indicate a $1.05R_J$ radius. It has a $6.6-8.2M_J$ minimum mass.

There is nothing preventing planets in the habitable zone from having stable orbits, although inner planets may have their orbit altered by the gas giant. The primary is moving off the main sequence, so a planet would have to orbit at 1.6AU or more to remain habitable through the next stage of stellar development.

Orbital simulations show that out of 100 test planets, seven are still in orbit after one million years.[79] Underwood, et al., report that a terrestrial planet orbit could remain in the habitable zone for the last one billion years, assuming the star is over two billion years old.[86]

HD 95128 (47 Ursa Majorais)

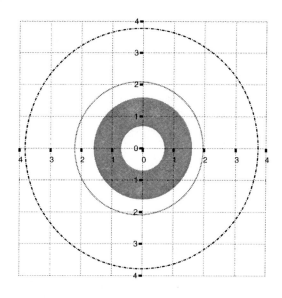

HD 95128 (47 Ursa Majorais) is located in the constellation Ursa Major, The Larger Bear, and is that constellation's 47th brightest star. It is type G0V star 45.9±3 light-years away (RA 10 59 27.97, D +40 21 48.92), with a 5.1 apparent magnitude.

47 UMa and the Sun are remarkably similar, their solar spectrums are almost identical. The effective temperature (5,731–5,954°K), absolute visual magnitude, and surface gravity are all consistent with its being a normal, old disk G0V main-sequence star. Its low chromospheric index and rotation speed of 16 days suggest an age of 4–8 Gyrs. It has a 1.05±.03$M_\odot$ mass, .06±.03 FeH metallicity ratio, and a 1.52$L_\odot$ luminosity.

The planet, the fourth found (1996, Lick Observatory[4]), is at the outer edge of the habitable zone, with a 2.09AU semi-major axis, a 1,090±3 day period, a .06±.06 eccentricity, and a 2.56±.02M_J minimum mass. Its radius is 1.1R_J, with an projected temperature of 194°K. There is less than 1% probability that this is a brown dwarf.

The second planet, the sixty-seventh discovered (2002 Lick Observatory[35]), has a 3.78±.05AU semi-major axis, a 2,640±2640 day period, a zero eccentricity, and a .76M_J min. mass estimate.

Simulations over one Gyr indicate stable orbits at .76, .8, .85, .9, .95, 1.05, 1.1, 1.15, 1.2, and 1.25AU, assuming the giant planets are at the minimum mass estimates or at 1.5 times the minimum masses.[80] Underwood, et al., report that a terrestrial planet orbit could remain in part of the habitable zone for the last one billion years.[86]

Coyote, a recent novel by Allen Steele, postulates a habitable terrestrial planet orbiting the first giant planet, with the radiated heat from the giant planet making up for the deficit inherent in being just beyond the habitable zone.

HD 75732 (55 Cancri, Rho¹ Cancri)

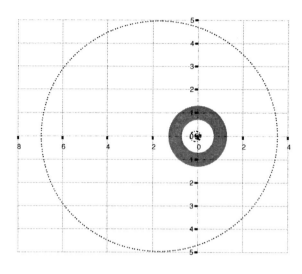

HD 75732 (55 Cancri, Rho Cancri) is the 55th brightest star in the constellation of Cancer, The Crab. It is a type G8V star about 40.862 (updated) light-years away (RA 08 52 37.6, D +28 20 02.6). It has a 5.95 apparent magnitude, a 5,279±62°K effective temperature (our Sun's is 5,778°K), a .33±.07 FeH ratio, and measurements indicate an age of 5 Gyr. It has a 36–42 day rotational period.

It has a .95±.08$M_\odot$ mass, and is a binary star with a separation of almost 1,100AU from a red type M4V star, apparent magnitude of 13.15, .13$M_\odot$ mass, .30$D_\odot$ diameter, and .0076$L_\odot$ luminosity.

The innermost planet, the 127th found (2004, Hobby-Eberly Telescope[65]), has a semi-major axis of .038±.001, a period of 2.808±.002 days, a .174±.127 eccentricity, and a .045±.01M_J minimum mass (14.21±2.95 M_{Earth}, or .824±.17$M_{Neptune}$).

The next planet, the fifth found (1996, Lick Observatory[5]), has a semi-major axis of .115±.003AU, a period of 14.67±.01 days, a .0197±.012 eccentricity, and .784±.09M minimum mass.

The middle planet, the seventy-eighth found (2002, Lick Observatory[41]), has a semi-major axis of .240±.008AU, a period of 43.93±.025 days, .44±.08 eccentricity, and a .217±.04M_J minimum mass.

The outer planet, the seventy-ninth found (2002, Lick Observatory[41]), has a semi-major axis of 5.257±.208AU, a period of 4517.4±77.8 days, a .327±.28 eccentricity, and a 3.912±.52M_J min. mass.

The three inner planets' masses do not affect the habitable zone. Orbital simulations show that out of 100 terrestrial test planets, fifty are still in orbit after one million years.[79] Planets form relatively easily (up to .6 Earth masses), and, in some cases, substantial water contents. Underwood, et al., report that a terrestrial planet orbit could remain in the habitable zone for the last one billion years.[86]

HD 120136 (Tau Bootis)

HD 120136 (Tau Bootis) is the 19th brightest star in the constellation Bootes, The Herdsman. It is a type F7V star about 48.9 light-years away (RA 13 47 17.34, D +17 27 22.31). It has an apparent magnitude of 4.5, a $1.3M_\odot$ mass, a 6,339±73°K effective temperature, a .23±.07 FeH ratio, and appears to be 1-2 Gyrs in age. It might be a variable star, according to SIMBAD.

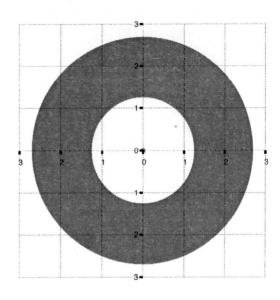

It is a main sequence star, with a rapid rotation of 4 days. The equatorial velocity is close to that of the measured velocity, induced by the planet orbiting it, suggesting that we are viewing the star nearly edge-on. Additionally, Tau Bootis is a binary star system with a separation of about 1,000AU. The second star is a red main sequence dwarf star of type M2, an 11.1 apparent magnitude, $.20M_\odot$ mass, $.58D_\odot$ diameter, $.0084L_\odot$ luminosity, a period around 2,000 years and an extremely high 0.91 orbital eccentricity.

The planet, the sixth found (1996, Lick Observatory[5]), has a semi-major axis of .047±.03AU, less than 5 million miles from the primary, and compared to the habitable zone has little impact on any orbits of planets inside that zone. Its minimum mass is $4.14\pm.06M_J$ with a period of 3.313±.001 days, and a low .04±.01 eccentricity.

One group has detected Tau Bootis' light reflected from the planet's atmosphere, yielding an orbital inclination of 29°, and a $8M_J$ mass.

Any terrestrial planets would be still be in the process of atmospheric reduction and building a stable crust. The likelihood of a breathable oxygen atmosphere is low. Orbital simulations show that out of 100 test planets, eighty are still in orbit after one million years (simulations of our solar system yields only eighty-one out of 100).[79] Underwood, et al., report that a terrestrial planet orbit could remain in the habitable zone for the last one billion years.[86]

HD 95735 (Lalande 21185)

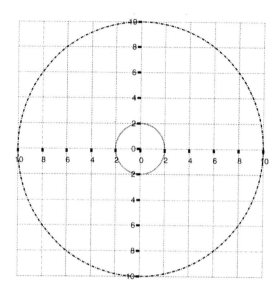

HD 95735 (Lalande 21185) is a small M2V type star in the constellation of Ursa Major, The Larger Bear, about 8.25 light-years away (RA 00 03 20.19, D +35 58 11.54). It has an apparent magnitude of 7.49, a mass of $.46 M_\odot$, but a $.006 L_\odot$ luminosity and a temperature of 3,300°K. This is the fourth closest system to our sun after Alpha Centauri, Barnard's Star, and Wolf 359.

The inner planet has a 2AU semi-major axis, a 5.8 year period, an eccentricity assumed to be zero (but actually unknown), and a $.9 M_J$ mass.

The outer planet has a 10AU semi-major axis, a 30 year period, an assumed zero eccentricity (but actually unknown), and a $1.6 M_J$ mass.

These two planets were the seventh and eighth planets announced (1996, Allegheny Observatory[6]), and were optically detected by measuring the "wobble" in the primary's path through space rather than using radial velocity measurements. There may be a third planet orbiting farther out, assuming that the mass of the second planet has not been underestimated.

The habitable zone is very close to the star, about 6.5 million to 7.2 million miles. At that distance, however, the rotation of a terrestrial planet would be tidally locked with the star. One side would have perpetual daylight and the other darkness after only a few million years. Such a planet would complete its orbit around the star in less than eleven days.

HD 9826 (Upsilon Andromedae)

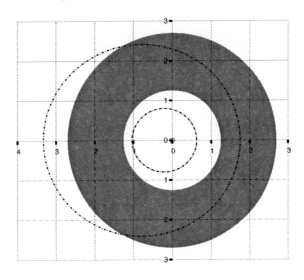

The fiftieth brightest star in the constellation Andromeda is HD 9826 (Upsilon Andromedae), a type F8V star about 43.9 light-years away (RA 01 36 48.52, D +41 24 38.71) with a 4.09 apparent magnitude, a $1.29M_\odot$ mass, a $6,212\pm64°K$ effective temperature, a $.13\pm.08$ FeH ratio, and is estimated to be 5 Gyrs in age, although its rapid 12-day rotation would seem to indicate an age of only 3 Gyrs.

A binary system main sequence star, Upsilon Andromedae is the second multi-planet system known, and the first to be discovered. The companion star (2MASSI J0136504+412332) has a spectral type of M4.5V and orbits about 750AU. Orbital eccentricity is unknown.

The first planet, the ninth discovered (1996, Lick Observatory[5]), has a .059AU semi-major axis, with a $4.617\pm.0003$ day period, a $.012\pm.15$ eccentricity, and a $.68\pm.01M_J$ minimum mass.

The second planet, the twenty-first found (1999, Lick Observatory, Whipple Observatory, Anglo-Australian Observatory, High Altitude Observatory[18,19]), has a $.828\pm.001$AU semi-major axis, a 241.3 ± 1.2 day period, a $.25\pm.13$ eccentricity, and a $1.94\pm.05M_J$ minimum mass.

The third planet, the twenty-second discovered (1999, Lick Observatory, Whipple Observatory, Anglo-Australian Observatory, High Altitude Observatory[18,19]), has a $2.54\pm.01$AU semi-major axis, a $1,299\pm30$ day period, a $.31\pm.11$ eccentricity, and a $4.02\pm.27M_J$ min. mass. This planet orbits across most of the habitable zone, making it impossible for stable orbits for terrestrial planets in that zone. Underwood, et al., report that a terrestrial planet orbit would not remain in the habitable zone for the last one billion years.[86] This planet could have a terrestrial-size habitable moon, and heat generated by the planet might keep the moon warm enough when it is outside the outer edge of the zone.

HD 186427 (16 Cygni B)

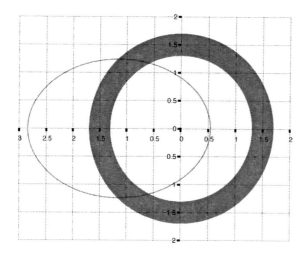

HD 186427 (16 Cygni B) is part of a triple star system in the constellation of Cygnus, The Swan. It is a type G3V star 69.7 light-years from Earth RA 19 41 51.97, D +50 31 03.08), with a 6.21 apparent magnitude, a $1.33L_\odot$ luminosity, a $1.19D_\odot$ diameter, a $5,772\pm25°K$ effective temperature (only 20° cooler than our Sun), a $.08\pm.04$ FeH ratio, mass of $.97–1.07M_\odot$, and a 29-day rotation. From its metallicity an age of 9 Gyrs is estimated. Its solar spectrum is identical to that of our sun, with a higher metallicity and luminosity.

The companion star, 16 Cyg A is about 843AU away. It's a type G1.5V, with a 5.96 apparent magnitude, a $1.68L_\odot$ luminosity, a $1.44D_\odot$ diameter, a 5,785°K effective temperature, a $1.05M_\odot$ mass, and a 26.9 day rotation period. Its orbital period is between 18,200 yrs and 1.3 Myrs, semi-major axis is between 877AU–15,180AU, and .54–.96 eccentricity.

16 Cygni A was observed at the same time as 16 Cyg B. No variations in its radial velocity measurements were detected. It can be assumed that no planets greater than about $.3M_J$ orbit 16 Cyg A.

16 Cyg C is a small red dwarf star that orbits 16 Cygni B. There has been insufficient observational time for any conclusions to be made as to its period or orbital parameters. If it is a low mass star, it is 80AU away; a higher mass of up to $.5M_\odot$ puts it at 150AU.

The planet, the tenth discovered (1996, McDonald Observatory[7]), has a $1.69\pm.03$AU semi-major axis, a 798.4 ± 6 day period, a $.68\pm.01$ eccentricity, and a $1.68\pm.18M_J$ minimum mass. Unfortunately, its orbit takes it through the habitable zone making any stable orbit impossible. Underwood, et al., concur with this conclusion.[86] Additionally, Terrestrial-sized moons around the planet would be outside the habitable zone too long to be habitable.

16 Cyg A (HD 186408) is compositionally identical to B, so its habitable zone should be clear.

HD 143761 (Rho Coronae Borealis)

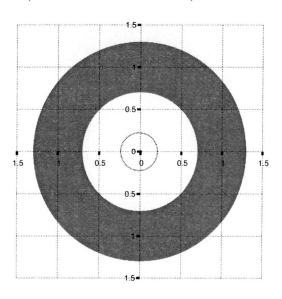

HD 143761 (Rho Coronae Borealis) is a G0V (maybe G2V) type star that is slightly bluer then our Sun and is 54.4 light-years away (RA 16 01 02.66, D +33 18 12.63). It is in the constellation of Corona Borealis, The Northern Crown, has a 5.42 apparent magnitude, a $1.77L_{\odot}$ luminosity (indicating that it is older than our Sun), a 5,853±25°K effective temperature, and a -.21±.04 FeH ratio. Its mass is about .95–1.05$M_{\odot}$, radius is 1.31$R_{\odot}$, and its rotation is 17 or 20 days. The approximate age of the star is 10–12.3 Gyrs. The inactivity of its chromsphere supports this age estimate. All this indicates an older Sun-like star near the end of its main-sequence life, possibly already starting hydrogen-shell burning.

The planet, the eleventh discovered (1997, High Altitude Observatory[8]), has a .224±.06AU semi-major axis, a 39.81±.24 day period, a .07±.15 eccentricity, and a minimum .99±.08M_J mass.

As for terrestrial planets, the diagram above is a blend of the projected habitable zone during normal main-sequence life (.8–1.6AU) and the habitable zone at the end of its main-sequence life (1.542–2.22AU). A terrestrial planet could orbit anywhere in the habitable zone without interference from a larger planet located within its orbit. The best orbit would be about 1.5AU. Orbital simulations show that out of 100 test planets, eighty are still in orbit after one million years.[79] Underwood, et al., report that a terrestrial planet orbit could remain in the habitable zone for the last one billion years.[86]

Ross 780 (Gliese 876)

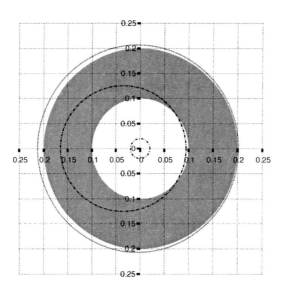

Ross 780 (Gliese 876) is one the few stars in this book without a Henry Draper number as an index. It is a small, dim, red, type M4V star, located about 15.32 light-years away in the constellation of Aquarius (RA 22 53 16.73, D-14 15 49.32), The Fishes, with a 10.16 apparent magnitude, a .32$M_\odot$ estimated mass, a .39$D_\odot$ diameter, a .002$L_\odot$ luminosity, a 3,100–3,250°K effective temperature, and a FeH ratio of zero, the same as our Sun.

This was the first discovery of a planet around an M dwarf by radial velocity measurements, and the only one out of 24 M-type stars studied over a period of four years.

The outer planet, the twelfth planet discovered (1998, Lick Observatory, Keck Observatory,[9] Haute-Provence Observatory, France, and European Southern Observatory, Chile[12]), has a .207AU semi-major axis, a 60.94 day period, a .0249 eccentricity, and a 1.935±.007M_J minimum mass.

The middle planet, the fifty-fifth planet (2000, Lick Observatory and Keck Observatory[33]) discovered, has a .13AU semi-major axis, a 30.12 day period, a .27 eccentricity, and a .56M_J min. mass.

The two giant planets are in a 2:1 orbital "resonance" about the primary that is remarkably stable.

The innermost planet, the 154th planet (2005, Lick Observatory[98]) discovered. It has a .0208067AU semi-major axis, a 1.93776 day period, and a .023±.003M_J (5.89±.054 Earth) mass.

Orbital simulations show that out of 100 test Earth-sized planets in the habitable zone, none are still in orbit after one million years.[79] Underwood, et al., concur, reporting that a terrestrial planet orbit would not remain in the habitable zone for the last one billion years.[86]

HD 145675 (Gliese 614, 14 Hercules)

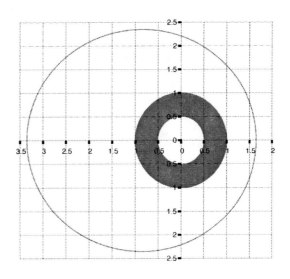

HD 145675 (Gliese 614, 14 Hercules) is a type K0V star 59.2±4 light-years away, the fourteenth brightest star in the constellation Hercules (RA 16 10 24.31, D +43 49 03.52). It has a 6.64 apparent magnitude, a .79–.90$M_\odot$ mass, a .88$D_\odot$ diameter, a 5,311±87°K effective temperature, a .43±.08 FeH ratio, and a .75$L_\odot$ luminosity. From its metallicity an age of 6 Gyrs is estimated.

The planet, the thirteenth found (1998, Haute-Provence Observatory, France[10]), has a 2.35±.05AU semi-major axis, a 1,796±8.3 day period, a .338±.011 eccentricity, and a 4.74±.06M_J min. mass. This planet is twice as close to 14 Hercules as Jupiter is to our Sun.

The primary star is close enough to us, and the planet is orbiting far enough out (14 arcseconds), that an attempt was made to visually detect the planet candidate. No images were obtained, eliminating the possibility that the planet was actually a brown dwarf in a more eccentric orbit than plotted.

The habitable zone is under the influence of the giant planet, but orbital simulations show that out of 100 test planets, twenty-six are still in orbit after one million years.[79] Only those with orbits under .7AU and a .2 or under eccentricity are stable. Underwood, et al., report that a terrestrial planet orbit could remain in part of the habitable zone for the last one billion years.[86]

HD 187123

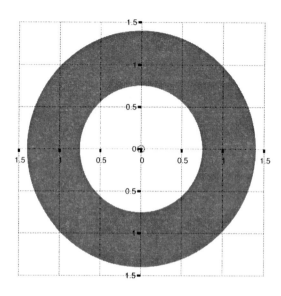

Located in the constellation of Cygnus, The Swan, is the fourteenth planet (1998, Keck Observatory and Anglo-Australian Observatory[11]) discovered (RA 19 46 58.11, D +34 25 10.28). HD 187123 is a type G5V star about 163 light-years away with a 7.86 apparent magnitude, a $1.35L_\odot$ luminosity, a 5,845±22°K effective temperature, and a .13±.03 FeH ratio. The star is very close to being spectroscopically identical to our sun (within 1% rms) with no enhancement of the iron line depths. Even though it is 35% brighter, it is remarkably similar to our Sun. Based on the close match, its mass is estimated at 1.04–1.05$M_\odot$, and from its metallicity an age of 5.5 Gyrs is estimated.

The planet has a .042AU semi-major axis, a 3.097 day period, a .03±.01 eccentricity (consistent with zero), and a .52±.03M_J min. mass.

Being so close to its primary, this planet has little influence on any possible planets in the habitable zone of the star.

Orbital simulations show that out of 100 test planets, eighty are still in orbit after one million years (simulations of our solar system yields only eighty-one out of 100).[79] Underwood, et al., report that a terrestrial planet orbit could remain in part of the habitable zone for the last one billion years.[86]

HD 210277

HD 210277 is officially a type G0V star about 71.72 light-years away in the constellation Aquarius (RA 22 09 29.49, D-.07 32 32.7), The Water Carrier, with a 6.63 apparent magnitude. However, the group that found the orbiting planet made measurements that give a stellar type of G7V, with a 5,532±28°K effective

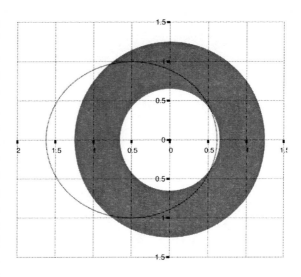

temperature, a .19±.04 FeH ratio, a .93$L_\odot$ luminosity, a .99±.05$M_\odot$ mass, and an age of 7–10 Gyrs (old, but not yet evolved into subgiant stage). From its metallicity an age of 8.5 Gyrs is estimated. Its rotation is estimated at 40.8 days.

The planet, the fifteenth found (1998, Keck Observatory and Anglo-Australian Observatory[11]), has a 1.12±.11AU semi-major axis, a 436±1.5 day period, a .45±.051 eccentricity, and a 1.29±.05M_J min. mass estimate.

No terrestrial planets could survive in any orbit that crosses a giant planet. If a terrestrial planet even comes within one astronomical unit of a giant planet its orbit is destabilized, except under special circumstances where the two planets reinforce each other's orbits.

Orbital simulations show that out of 100 test planets, none are still in orbit after one million years.[79] Underwood, et al., report that a terrestrial planet orbit would not remain in the habitable zone for the last one billion years.[86]

However, a terrestrial-sized moon might exist. Such a moon in this system could be habitable, despite that portion of the giant planet's orbit outside the habitable zone. Gas giant planets on the order of Jupiter's size are generating heat in their own right, which could provide the extra energy needed. The winters would be harsh and the summers boiling. Part of the time, the giant planet would be shading the moon from the harshest part of the summer while providing extra warmth during the winters.

HD 195019

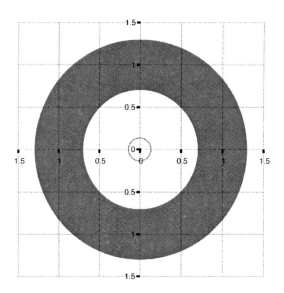

HD 195019 is a type G3IV/V star in the constellation of Delphinus, The Dolphin. It is 121.8 light-years away and has a 6.91 apparent magnitude (RA 20 28 18.63, D +18 46 10.18). Based on its spectral type, it is assumed to have a 1.02±.05$M_\odot$ mass. It has an effective temperature of 5,842°K, although it could be 5,700°K if measurements are polluted by its stellar companion, a .09±.04 FeH ratio, and a 22-day rotation. Based on the low chromospheric activity and rotation its age is estimated at 9.5 Gyrs. It is probably slightly evolved.

HD 195019 is also a binary star. The companion, BD +18°4505C (ADS13886B), is 3 magnitudes fainter, 9.8, and currently separated from the primary by 150AU. Not much more is known about this star.

The planet, the sixteenth discovered (1998, Lick Observatory and Keck Observatory[12]), has a .136±.004 semi-major axis, an 18.2 days period, a .02±.02 eccentricity, and a 3.55±.05M_J minimum mass.

Terrestrial planets could exist anywhere in the habitable zone without problems induced by the giant planet.

Orbital simulations show that out of 100 test planets, seventy-three are still in orbit after one million years.[79] Underwood, et al., report that a terrestrial planet orbit could remain in the habitable zone for the last one billion years.[86]

The primary is apparently entering its post-main sequence life and should be increasing its luminosity to double, or higher, that of our sun. The companion, although 150 AU away would still be almost forty times as bright (-16.4) as our moon (-12.5) when at full brightness, but still almost 1,000 times dimmer than our sun.

HD 217107

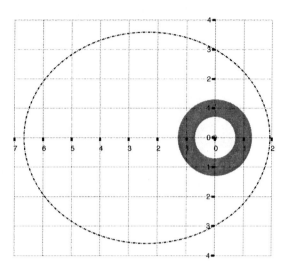

HD 217107 is a type G8IV star located some 64.25 light-years from Earth in the constellation Pisces, The Fishes, with a 6.17 apparent magnitude (RA 22 58 15.45, D-02 23 42.4). Normally such a star would have a mass estimate of $.98\pm.04M_\odot$, but measurements of the metallicity ($.37\pm.05$ FeH ratio) indicate it should have a $.98M_\odot$ mass. It has a 39-day rotation, and its age is estimated at 9.5–9.9 Gyrs. Its effective temperature is $5,646\pm34°K$ (our Sun's is $5,778°K$).

Like most of the previously discovered planets, this one, the seventeenth discovered (1998, Lick Observatory and Keck Observatory[12]), is close-in with a $.074\pm.002$AU semi-major axis, a $7.1269\pm.00022$ day period, a $.13\pm.02$ eccentricity, and a $1.37\pm.14M_J$ minimum mass.

If the orbital inclination were high enough for this to be a stellar object, its mass (over $20M_J$) would be sufficient for its tidal influences to have sped up the rotation of the star's chromosphere to match its orbital period, 7.13 days. The fact that they are different indicates that this is a planet and not a brown dwarf star.

The outer planet, the 158th discovered (2005, Keck Observatory[99]), has a 4.3 ± 2AU semi-major axis, a period of $3,150\pm1,000$ days, a $.55\pm.2$ eccentricity, and a $2.1\pm1M_J$ mass.

The habitable zone is fairly clear of the gravitational influence of the gas giant, almost any projected orbit in the zone would work.

Orbital simulations show that out of 100 test planets, seventy-five are still in orbit after one million years.[79] Underwood, et al., report that a terrestrial planet orbit could remain in the habitable zone for the last one billion years.[86]

HD 13445 (Gliese 86)

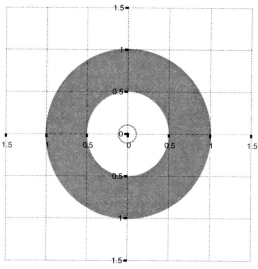

HD 13445 (Gliese 86), is a type K1V star 35.86 light-years from Earth in the constellation Eridanus, The River (RA 02 10 14.42 D -50 50 00.5). It has an apparent magnitude of 6.17, a 5,163±37°K effective temperature, a -.24±.05 FeH ratio, a .79$M_\odot$ mass, a .27$L_\odot$ luminosity, and a 31-day rotation. From its metallicity an age of 3.2 Gyrs is estimated.

The planet, the eighteenth found (1999, La Silla, European Southern Observatory, Chile[13]), has a .11±.007AU semi-major axis, a period of 15.78±.02 days, a .04 eccentricity, and a 4±.02M_J estimated minimum mass.

There is a brown dwarf in this system as well (1999, La Silla, European Southern Observatory, Chile[14]). It appears to be 16–20AU from the primary. No orbital parameters are available. An approximate spectral type for Gliese 86B is at the transition from L to T dwarfs, also called "early T dwarf," assuming the classification by Leggett et al.[15] (2000). Although present brown dwarf evolutionary models do not cover the mass and age range of these objects, a 70 M_J upper limit of the mass is inferred, but it could be as low as a 15M_J mass.

The Brown Dwarf in this system is not a factor, unless it is in an eccentric orbit that brings it to within 7AU of the habitable zone. The gas giant is in a nearly circular orbit, so its influence should not be great. The farther the terrestrial planet is from the gas giant, the more stable its orbit will be.

Orbital simulations show that out of 100 test planets, eighty-four are still in orbit after one million years (simulations of our solar system yields only eighty-one out of 100).[79] Underwood, et al., report that a terrestrial planet orbit could remain in the habitable zone for the last one billion years.[86]

HD 168443

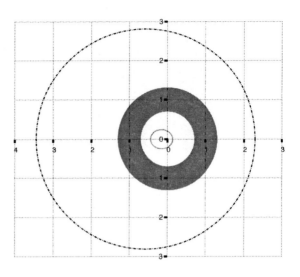

HD 168443 is a type G8IV star 68 light-years away in the constellation of Serpens, The Serpent (RA 18 20 04.11, D-09 35 34.6), with a 6.91 apparent magnitude, a 5,617±35°K effective temperature, and a 2.1$L_\odot$ luminosity. Given its apparent magnitude and distance, its absolute magnitude is 5.03, about 1.5 magnitudes higher than indicated for its temperature. This places it in the subgiant status and spectral type. HD 168443 is similar to 70 Vir. in mass, surface characteristics, and metallicity. The rotation is 37 days, giving an implied age of 7.8 Gyrs. The subgiant status would place it in the age of 7–10 Gyrs, but that status is tentative. From its metallicity (06±.05 FeH) an age of 2.6 Gyrs is estimated. Its mass has been estimated at .84$M_\odot$ and at 1.05$M_\odot$, hence a 1.01$M_\odot$ compromise.

The inner planet, the nineteenth found (1999, Keck Observatory[16]), has a .295AU semi-major axis, a 57.8 day period, a .53 eccentricity, and a 5.04M_J minimum mass.

The outer object, (2001, Keck Observatory[16]) has a 2.87AU semi-major axis, a 1,770±.25 day period, a .20±.02 eccentricity, and a 16.96±.027M_J min. mass. It is apparent it is a type T brown dwarf.

A terrestrial planet inside the habitable zone is probably impossible. The brown dwarf, even at the lower end of its mass limit, would still exert too much gravitational influence on any bodies inside the habitable zone. Any planet inside the zone would find itself ejected from the system within a million years, if it had time to form.

Orbital simulations show that out of 100 test planets, none are still in orbit after one million years.[79] Underwood, et al., report that a terrestrial planet orbit would not remain in the habitable zone for the last one billion years.[86]

HD 75289

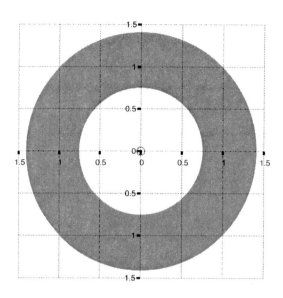

HD 75289 is G0V type star 153 light-years from Earth light-years away in the constellation Vela, The Sails (RA 08 47 40.38, D -41 44 12.45). It has a 6.35 apparent magnitude, a $1.99L_\odot$ luminosity, a 6,143±53°K effective temperature, a .28±.07 FeH ratio, a $1.1±.05M_\odot$ mass, a $1.4R_\odot$ radius, a 16-day rotation, and a 5.6 Gyr estimated age. Its mass is slightly higher than the typical G0 dwarf because of its high metallicity. It has a low chromospheric activity level.

A low-mass stellar companion is located approximately 621±10 AU from the primary. the spectrum and color of the companion is consistent with it being an M2 to M5 type star. The observation that detected this star rules out additional stellar compnaions beyond 140AU and substellar companions with $.05M_\odot$ masses or greater between 400 and 2,000AU.

The planet, the twentieth discovered (1999, La Silla, European Southern Observatory, Chile[17]), has a .047±.003AU semi-major axis, a 3.5±.009 day period, an eccentricity of zero or nearly zero, and a $.42±.01M_J$ minimum mass (1.4 times the mass of Saturn). The planet has a 1,260°K estimated temperature (water boils at 373°K).

Because of its close orbit to its primary there is a good chance that the planet is orbiting close to the plane of the star's equator, thus making it a candidate for transit detection.

As close as the planet is to its primary, there are stable orbits in the habitable zone. Orbital simulations show that out of 100 test planets, eighty-four are still in orbit after one million years (simulations of our solar system yields only eighty-one out of 100).[79] Underwood, et al., report that a terrestrial planet orbit could remain in the habitable zone for the last one billion years.[86]

HD 17051 (Iota Horologii, HR 810)

HD 17051 (Iota Horologii) is the ninth brightest star in the constellation Horologium, The Clock (RA 02 42 31.65, D -50 48 12.3). It is 50.5 light-years away, a type G0V, has a 5.4 apparent magnitude, a 1.52L$_\odot$ luminosity, a 6,252±53°K effective temperature, a .26±.06 FeH ratio, and a 7.9 to 8.6-day rotation. The age is estimated at between 30 Myrs and 2 Gyrs, although one

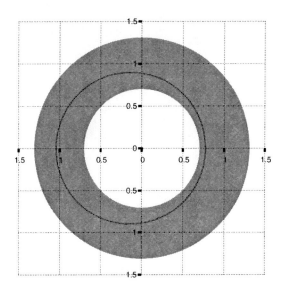

measurement gives an age of 6.89 Gyrs. The most probable is 1.56 Gyrs, putting this star in the zero-age main sequence stage. The mass has not been established but was assumed to be 1.03M$_\odot$ for the purpose of establishing the mass of the planet. The chromosphere of the star is very active.

The planet, the twenty-third found (1999, La Silla, European Southern Observatory[20]), has a .909±.104AU semi-major-axis, a 312±10 day period, a .15±.08 eccentricity, and a 2.25±.18M$_J$ minimum mass. There are upper limits of 5.18M$_J$, 7.24M$_J$ (min. 3.2M$_J$), and 16.0M$_J$ (min. 7.08M$_J$), with confidence levels that these are the true maximums/minimums of 90%, 95%, and 99%. The planet's closest approach is just outside the Sun-Venus orbit and goes out to the Sun-Earth orbit.

The planet's orbit precludes any chance of a stable orbit for a terrestrial planet in the habitable zone. However, the opportunity for a terrestrial-sized moon, or even several, is easily believable, especially if the maximums approach Brown Dwarf status. The only problem would be in placing them outside of the radiation belts such a planet would have (inside would make them tidally locked with one side always facing the planet). Such moons would be geologically active.

Orbital simulations show that out of 100 test planets, none are still in orbit after one million years.[79] Underwood, et al., report that a terrestrial planet orbit could be confined to part of the habitable zone for the last one billion years.[86]

HD 130322

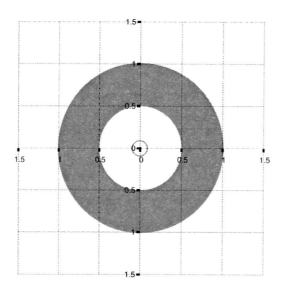

HD 130322 is a type K0V star in the constellation Virgo, The Virgin, at a distance of 97.8 light-years (RA 14 47 32.73, D -00 16 53.3). It has an 8.04 apparent magnitude, a 5,392°K effective temperature, .03±.04 FeH ratio, a .5L$_\odot$ luminosity, a .79±.25M$_\odot$ mass, a .85R$_\odot$ radius, a 8.7-day rotation, and the age is estimated at 350 Myrs. The age makes the star fairly active, introducing a higher level of uncertainty in the numbers derived for the planet.

The planet, the twenty-fourth discovered (1999, La Silla, European Southern Observatory, Chile[17]), has a .092±.04AU semi-major axis, an 8.7 day period, a .05±.002 eccentricity, and a 1.15±.07M$_J$ minimum mass, and a 6.8M$_J$ maximum mass (assuming, like HD 75289, that the planetary orbit is reasonably coplanar with the primary's equator). It has a 720°K temperature (water boils at 373°K).

As close as the planet is to its primary, stable orbits are possible in the habitable zone. Orbital simulations show that out of 100 test planets, eighty-four are still in orbit after one million years (simulations of our solar system yields only eighty-one out of 100).[79] Underwood, et al., report that a terrestrial planet orbit could remain in the habitable zone for the last one billion years.[86]

HD 192263

HD 192263 is in the constellation Aquilla, the Eagle, and is a K0V type at a distance of 64.87 light-years (RA 20 13 59.84, D -00 52 00.75). It has a 7.79 apparent magnitude, a .69-.75M$_\odot$ mass, and a 4,947°K effective temperature. From its metallicity (-.02±.06 FeH ratio) an age of 300 Myrs is estimated. The rotation should be 8 days, but the chromosphere is very active.

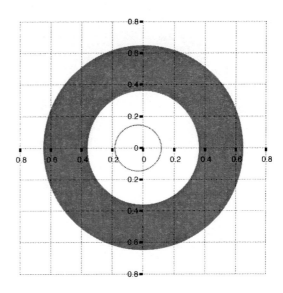

The velocity spectra seems to indicate a rotation of 26.7 days, uncomfortably close to the period of the planet. The false alarm probability for this is 4.7%, suggesting the peak may be coincidental. Given the short 8-day rotation implied by the chromospherical activity and the broadening of spectra that gives a low velocity Doppler, the star is probably being viewed within 30° of the pole. This increases the probable mass of the planet significantly. Also, the residuals of the measurements are 4.5 m/s instead of the higher jitter of 10 m/s that the active chromosphere would indicate should be the case.

Assuming the measurements are correct, the planet, the twenty-fifth discovered (1999, Keck Observatory[21]), has a .15AU semi-major axis, a 23.87±.14 day period, an eccentricity that could be either .03 or .22, a .78M$_J$ minimum mass, and a 486°K temperature (water boils at 373°K). Internal heating could raise the temperature another 10–20°K. The chart above assumes a .22 eccentricity.

Underwood, et al., report that a terrestrial planet orbit could remain in the habitable zone for the last one billion years.[86]

HD 209458

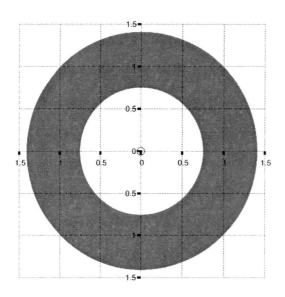

HD 209458 is a type G0V star 153 light-years away in the constellation Pegasus (RA 22 03 10.7, D +18 53 04). It has a 1.03–1.15$M_\odot$ mass, a 1.15$R_\odot$ radius, a 6,117±26°K effective temperature, and a .02±.03 FeH ratio. It is chromospherically inactive and estimated to be 4.5Gyrs old. It has at least a 17-day rotation.

The planet, the twenty-sixth discovered (1999, Keck Observatory[22]), has a .046±.01AU semi-major axis, a 3.52474541±.00000025 day period, a zero eccentricity, .02±.02, a .69±.05M_J minimum mass, and a 1,130±150°K effective temperature. This is a firm number, as it has been observed actually transiting the primary. Based on the observed transits, the radius of the planet is 1.31R_J (±.5), the minimum mass is .69±.05M_J, and the density is .27 g/cm^3 (±.04) (Saturn has a density of .69 g/cm^3).

The planet, although less massive than Jupiter, is larger. This is due to the heat absorption because of its close orbit. For comparison, a Jupiter mass planet, if it were at a distance of .05AU (4.5 million miles), would have a radius of 1.5R_J, consistent with the observed size of HD 75289b. The orbital inclination is less than 30° from the primary's equator. The planet is losing 10,100 tons of atmosphere per second as its sun blows the molecules away. There is a comet-like tail of molecules stretching tens of millions of miles from the planet. Even so, the planet will likely survive until the star goes nova.

As close as the planet is to its primary, stable orbits are possible in the habitable zone. Orbital simulations show that out of 100 test planets, 82 are still in orbit after one million years (simulations of our solar system yields only eighty-one out of 100).[79] Underwood, et al., report that a terrestrial planet orbit could remain in the habitable zone for the last one billion years.[86]

HD 10697

HD 10697 is a type G5IV star in the constellation Andromeda at a distance of 97.8 light-years (RA 01 44 55.82, D +20 04 59.33). It has a 6.29 apparent magnitude, a $1.10M_\odot$ mass, a 5,641°K effective temperature, a .14±.04 FeH ratio, and it is a slowly rotating star that is chromospherically inactive.

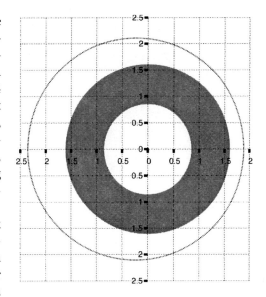

The planet (1999, Keck Observatory[21]) has a semi-major axis of 2.12AU, with a 1,074±7 day period, a near zero eccentricity, and a 6.08±.03M_J minimum mass. It has a 264°K estimated temperature (water freezes at 273°K). Internal heating could raise the temperature another 10–20°K.

The gas giant is too large for stable orbits in the habitable zone. The planet's gravitational pull would prevent any material from clumping together enough to form a planet. Jupiter, in our system, similarly is thought to have prevented a planet from forming between Jupiter and Mars.

Orbital simulations show that out of 100 test planets, none are still in orbit after one million years.[79] Underwood, et al., report that a terrestrial planet orbit would not remain in the habitable zone for the last one billion years.[86]

HD 37124

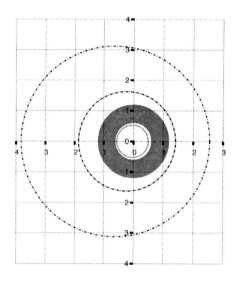

HD 37124 is a type G4V star in the constellation Taurus, The Bull, and is 107.5 light-years away (RA 05 37 02.48, D +20 43 50.83). It is a slowly rotating star that is chromospherically inactive. It has 5.07 apparent magnitude, a $.91\pm.15M_\odot$ mass, a 5,546°K effective temperature, a -.38±.04 FeH ratio, and a $.82L_\odot$ luminosity. From its metallicity an age of 3.8 Gyrs is estimated.

The inner planet, the twenty-eighth discovered (1999, Keck Observatory[21]), has a .53±.003AU semi-major axis, a 152.46±.369 day period, a .055 eccentricity, and a .61±.04M_J minimum mass. The planet is expected to have a temperature of 327°K (water boils at 373°K). Internal heating could raise the temperature another 10–20°K.

The middle planet, the eighty-first discovered (2002, Keck Observatory[43]), in the system with a 1.64AU semi-major axis, 843.6 day period, .6M_J mass, and a .14 eccentricity.

The outer planet, the 159th discovered (2005, Keck Observatory[99]), in the system with a 3.19AU semi-major axis, 2,295 day period, .66M_J mass, and a .2 eccentricity.

Orbital simulations, based on the two-planet model previously believed to be accurate, show that out of 100 test planets, none are still in orbit after one million years.[79] Underwood, et al., concur, and report that a terrestrial planet orbit would not remain in the habitable zone for the last one billion years.[86]

With three planets, the inner two planets now bracket the habitable zone, but are probably still to close to it to allow stable orbits for terrestrial-sized planets.

HD 134987

HD 134987 is a type G5V star in the constellation Libra, The Scales, at a distance of 81.5 light-years (RA 15 13 28.6676, D -25 18 33.649). It has a 6.45 apparent magnitude, a 1.05$M_\odot$ mass, and a 1.51$M_\odot$ luminosity, a 5,776±29°K effective temperature, and a .30±.04 FeH ratio. It has low chromospheric activity. From its metallicity an age of 6 Gyrs is estimated.

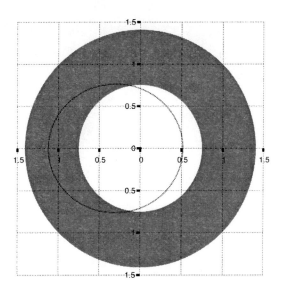

The planet, the twenty-ninth detection (1999, Keck Observatory[21]), has a .821±.01AU semi-major axis, a 265±5.3 day period, .24–.37 eccentricity (the plot above assumes .37), a 1.63±.02M_J minimum mass, and a 315°K planetary temperature. Internal heating could raise the temperature another 10–20°K.

The planet's orbit through the habitable zone substantially decreases the chances of a stable orbit for a terrestrial planet in that zone. Any moons would be too hot to be habitable, especially considering the additional heat added by the planet itself.

Orbital simulations show that out of 100 test planets, none are still in orbit after one million years.[79] Underwood, et al., report that a terrestrial planet orbit could remain in only part of the habitable zone for the last one billion years.[86]

HD 177830

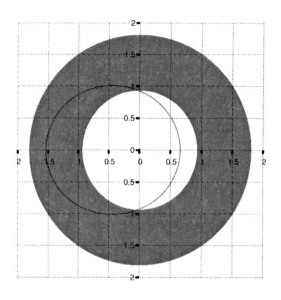

HD 177830 is an evolved subgiant of the type K0IV with photometry placing it close to giant status. It is in the constellation of Lyra, The Lyre, at a distance of 192 light-years (RA 19 05 20.7735, D +25 55 14.379). It has a 7.18 apparent magnitude, a $1.17\pm.14M_\odot$ mass, a $3.99M_\odot$ luminosity, a 4,804°K effective temperature, and a .33±.09 FeH ratio. It is chromospherically inactive. Evolved subgiants have complex variability periods that range from less than a day to several hundred days. The detected planet could be merely a component of the rotating chromosphere of the star. More study is required.

The planet, the thirtieth discovered (1999, Keck Observatory[21]), has a 1.1±.1AU semi-major axis, a 391±18 day period, a .4±.31 eccentricity, and a $1.24\pm.24M_J$ minimum mass. The temperature is 362°K. Internal heating could raise the temperature another 10–20°K.

Cutting through the habitable zone for most of its orbit, this planet makes it very difficult for a terrestrial planet to survive. Currently, any moons would be too hot to be habitable, especially considering the additional heat added by the planet itself.

Orbital simulations show that out of 100 test planets, one is still in orbit after one million years.[79] Underwood, et al., report that a terrestrial planet orbit could be confined only to part of the habitable zone for the last one billion years.[86]

HD 222582

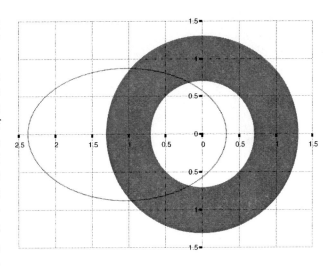

HD 222582 is a type G5V star in the constellation Aquarius, The Water Carrier, at a distance of 136.9 light-years (RA 23 41 51.5299, D -05 59 08.726). It has a 7.7 apparent magnitude, a $1.0 \pm .02 M_\odot$ mass, a $1.25 L_\odot$ luminosity. Measurements indicate that this star's temperature (5,843±38°K) and makeup (.05±.05 FeH ratio) are quite similar to our Sun, although the chromosphere appears to have less magnetic activity than the Sun. From its metallicity an age of 5.7 Gyrs is estimated.

The planet, the thirty-first found (1999, Keck Observatory[21]), has a 1.35AU semi-major axis, a 577±1.4 day period, a .76±.01 eccentricity, and a $5.2 \pm .02 M_J$ minimum mass. The temperature of the planet should be 234°K, although internal heating will raise this by 10–20°K (water freezes at 273°K).

Dropping through the habitable zone twice every orbit makes stable orbits for other planets in the habitable zone impossible. With its large eccentricity, not even terrestrial moons of the planet would be inhabitable.

Orbital simulations show that out of 100 test planets, none are still in orbit after one million years.[79] Underwood, et al., report that a terrestrial planet orbit would not remain in the habitable zone for the last one billion years.[86]

HD 1237 (Gleise 3021)

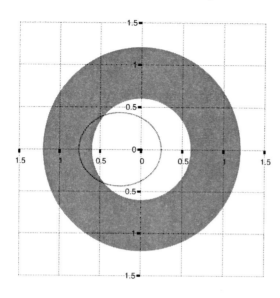

HD 1237 (Gleise 3021) is a type G6V star in the constellation Hydrus, The Water Serpent, and is 75.4 light-years away (RA 00 16 12.6775, D -79 51 04.254). The star has a 6.59 apparent magnitude, a $.93M_\odot$ mass, a $.94R_\odot$ radius, a 5,536°K effective temperature, a $.12\pm.06$ FeH ratio, and a $.66L_\odot$ luminosity. It is chromospherically active enough to give an age of 600 million years. This is supported by a fast rotation and an X-ray luminosity much too high for an old type G dwarf.

The planet, the thirty-second discovered (2000, La Silla, European Southern Observatory, Chile[23]), has a .505AU semi-major axis, a $133.8\pm.02$ day period, a $.51\pm.02$ eccentricity, and a $3.45\pm.13M_J$ minimum mass. With an orbit ranging from .25 AU to .75AU, the planet's temperature ranges from 440°K at periastron (closest to star) to 260°K at apastron (furthest from star).

Cutting through the habitable zone inner edge, this planet prevents any stable terrestrial planet orbit except at the outer edges. In addition, with such a high range of temperatures (very hot at apastron, well above boiling at periastron), any terrestrial moons would probably be uninhabitable.

Orbital simulations show that out of 100 test planets, none are still in orbit after one million years.[79] Underwood, et al., report that a terrestrial planet orbit would not remain in the habitable zone for the last one billion years, primarily because it isn't that old yet! However, they do conclude that a terrestrial planet in the outer portion of the habitable zone would be stable in a billion-year time-period, especially as the star gets hotter when as it ages.[86]

HD 89744

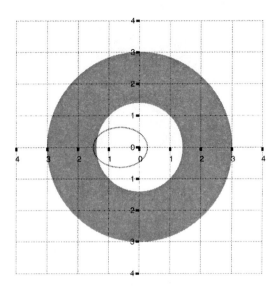

HD 89744 is a type F7V star in the constellation Ursa Major, The Larger Bear, and is 130.4 light-years away (RA 10 22 10.5621, D +41 13 46.308). The star has a 5.76 apparent magnitude, a $1.40M_\odot$ mass, a $2.14R_\odot$ radius, a $6.57L_\odot$ luminosity, a 6,234°K effective temperature, a .22±.05 FeH ratio, and an age of 2.0 Gyrs. It has a 9-day rotational period, supporting the estimate of a young age. It is also not photometrically stable. Its light output appears to vary at irregular intervals, perhaps due to features such as sunspots. The Hipparcos catalog lists it as a micro-variable star.

The planet, the thirty-third discovered (2000, Whipple Observatory and Lick Observatory[24]), has a .883±.007AU semi-major axis, a 256±.734 day period, a .7±.02 eccentricity, and a $7.17±.57M_J$ minimum mass. Based on the rotation, radius, and broadening of the velocity measurements of the primary, an orbital inclination of 42° from the primary's equator is implied. If the orbit were coplanar, the mass of the planet would be $10.8M_J$, large, but below the brown dwarf cutoff limit.

There is an L8V brown dwarf at a distance of 100AU, its eccentricity is unknown.

Orbital simulations show that out of 100 test planets, none are still in orbit after one million years.[79] Underwood, et al., report that a terrestrial planet orbit could remain in the outer portion of the habitable zone for the last one billion years.[86]

HD 12661

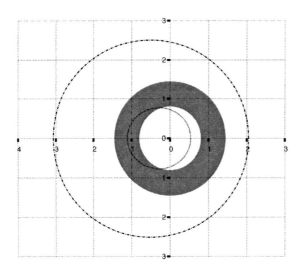

HD 12661 is usually listed in references (SIMBAD, Hipparcos) as a K0 star; however, the color index (apparent visual magnitude of 7.43 compared to its blue magnitude of 8.16) is more consistent with a G6V star. It is in the constellation Ara, The Altar, and is 121.1 light-years away (RA 02 04 34.2879, D +25 24 51.499). It has a 1.07±.02$M_\odot$ mass, a 1.096$R_\odot$ radius, a 5,702±36°K effective temperature, a .36±.05 FeH ratio, a 36-day rotation, and is chromospherically quiet.

The inner planet, the thirty-fourth discovered (2000, Lick Observatory and Keck Observatory[25]), has a .823AU semi-major axis, a 263.3±1.2 day period, a .35±.03 eccentricity, and a 2.3M_J minimum mass.

The outer planet, the eightieth planet discovered (2003 Lick Observatory, Keck Observatory[42]), has a 2.56AU semi-major axis, a 1,444.5±12.5 day period, a .2±.04 eccentricity, and a 1.56±.01M_J min. mass.

The inner planet dominates the habitable zone. Terrestrial moons would be difficult because just under half their orbit would be below the habitable zone. Balancing that is the fact that the moon would be in the planet's shadow for about 40% of that time.

Orbital simulations show that out of 100 test planets, none are still in orbit after one million years.[79] Underwood, et al., report that a terrestrial planet orbit would not remain in the habitable zone for the last one billion years.[86]

HD 16141 (79 Ceti)

HD 16141 (79 Ceti) is a type G5IV star in the constellation Cetus, The Whale, 117 light-years from the Earth (RA 02 35 19.9283, D -03 33 38.167). It has a 6.83 apparent magnitude, almost a full magnitude above the zero-age main-sequence. It has a 1.01$M_\odot$ mass, a 2.08$L_\odot$ luminosity, a 5,801±30°K effective temperature, a .15±.04 FeH ratio, and a 1.01$D_\odot$ diameter. It is rotating slowly, and is chromospherically inactive. As the star is unusually bright for its spectral type despite a cooler surface temperature, it is classified as a subgiant star that has begun to evolve off the main sequence and it may soon be unable to sustain hydrogen burning at its core. From its metallicity (.15±.04 FeH ratio) an age of 6.6 gigayears is estimated.

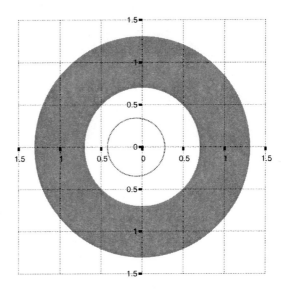

The planet, the thirty-fifth discovered (2000, Keck Observatory[26]), has a .35AU semi-major axis, a 75.8±.4 day period, a .21±.7 eccentricity, and a .22±.01M_J min. mass (Saturn has a mass of .32M_J).

The planet is not quite far enough away from the habitable zone to leave it completely unscathed. Stable orbits need to be above 1AU, preferably at the outer edges of the habitable zone at 1.4AU — within the orbit of Mars in our Solar System — with an orbital period of 611 days (or 1.67 years).

Orbital simulations show that out of 100 test planets, twenty-eight are still in orbit after one million years.[79] Underwood, et al., report that a terrestrial planet orbit could remain in the habitable zone for the last one billion years.[86]

HD 46375

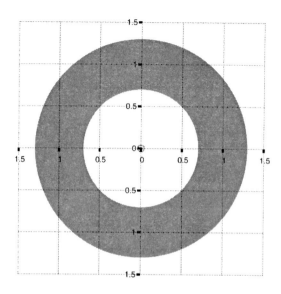

HD 46375 is a type KIV-V star in the constellation of Monoceros, The Unicorn, approximately 108.8 light-years away (RA 06 33 12.6237, D +05 27 46.532). It has a 7.84 apparent magnitude, almost a full magnitude above the zero-age main sequence. It has a $1.00\pm.17M_\odot$ mass, a $5,268\pm55°K$ effective temperature, a $.20\pm.06$ FeH ratio, is rotating slowly, and is chromospherically inactive. From its metallicity an age of 4.5 Gyrs is estimated.

The planet, the thirty-sixth detected (2000, Keck Observatory[26]), has a .04±.001AU semi-major axis, a 3.024 day period, a .02±.02 eccentricity, nearly zero, and a $.25\pm.01M_J$ minimum mass (Saturn has a mass of $.32M_J$). The planet is so close, it should "spin up" the primary's chromosphere rotation. Because it hasn't, the planet has an upper limit of 15 M_J. The planet has a 1,500°K temperature.

The gas giant exerts little to no gravitational effect on any possible orbits in the habitable zone. Orbital simulations show that out of 100 test planets, eighty-five are still in orbit after one million years (simulations of our solar system yields only eighty-one out of 100).[79] Underwood, et al., report that a terrestrial planet orbit could remain in the habitable zone for the last one billion years.[86]

HD 52265

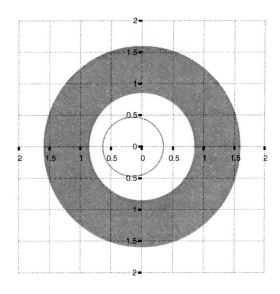

HD 52265 is a type G0V star, based on the distance measured by Hipparcos and a subsequent 4.06 absolute magnitude. This is in contrast to the Bright Star Catalog and SIMBAD classi-fication of G0III-IV. It is in the constellation Crater, The Cup, at a distance of 91.2 light-years (RA 07 00 18.0363, D -05 22 01.783). HD 52265 has a 6.3 apparent magnitude, a $1.13M_\odot$ mass, a $1.98L_\odot$ luminosity, a 6,103±52°K effective temperature, a .23±.07 FeH ratio, a $1.1R_\odot$ diameter, and the chromosphere is quiet. Its metallicity is 11% higher than our Sun. From its metallicity an age of 3.5–4 Gyrs is estimated. Thus, this star appears to be a slightly evolved metal-rich G0 dwarf.

The planet, the thirty-seventh detected (2000, Keck Observatory[27]), has a .49±.008 semi-major axis, a 119±.1 day period, a .29±.09 eccentricity, and a 1.00±.19M_J minimum. mass.

The planet's gravity influence extends into the habitable zone, making it difficult, but not impossible for stable terrestrial planetary orbits.

Orbital simulations show that out of 100 test planets, twenty-two are still in orbit after one million years.[79] Underwood, et al., report that a terrestrial planet orbit could remain in the habitable zone for the last one billion years.[86]

BD -10°3166

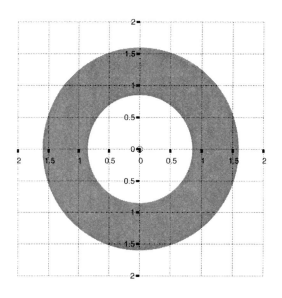

BD -10°3166 is a type K0V star in the constellation Crater, The Cup. There is no published parallax, so the distance to this star is unknown (RA 10 58 28.780, D -10 46 13.39). It has a 5,320°K effective temperature, a .33 FeH ratio, a 10.0 apparent magnitude, and is a main sequence star. If it were a subgiant it would have to be over 600 light-years away to have its magnitude, and have a velocity in excess of 180 km/sec, making it a member of the galactic halo. However, its metallicity is far too high for it to be a member of that group. Far more likely is a velocity of 26 km/sec. Its metallicity and temperature give a $1.1M_\odot$ estimated mass, and age of 4 Gyrs. The chromosphere is slowly rotating and inactive.

The planet, the thirty-eighth detected (2000, Keck Observatory[27]), has a .046AU semi-major axis, with a 3.488±1 day period, a .06 eccentricity, and a $.48M_J$ minimim mass.

The gas giant exerts little to no gravitational effect on any possible orbits in the habitable zone. Orbital simulations show that out of 100 test planets, eighty-nine are still in orbit after one million years (simulations of our solar system yields only eighty-one out of 100).[79] Underwood, et al., report that a terrestrial planet orbit could remain in the habitable zone for the last one billion years.[86]

HD 82943

HD 82943 is a type G0 star in the constellation, Hydra, The Water Snake, at a distance of 89.52 light-years (RA 09 34 50.7361, D -12 07 46.365. It has a 6.5 apparent magnitude, a 1.2$M_\odot$ mass, a 6,016±30°K effective temperature, and a .30±.04 FeH ratio. From its metallicity an age of 600 million years is estimated, although standard references

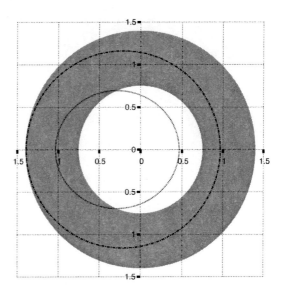

give a 4.7 Gyrs age estimate. A fair amount of the easy-to-burn element Lithium has been detected, suggesting a planet engulfment after the star reached main sequence. The not-too-deep convective zone of this G0 dwarf allows for the survival of this element.

The outer planet, the thirty-ninth found (2000, La Silla, European Southern Observatory, Chile[28]), has a 1.18AU semi-major axis of, a 435.1 day period, a .18 ecc., and a 1.84M_J mini. mass.

The inner planet, the sixty-second discovered (2001, La Silla, European Southern Observatory, Chile[34]), has a .75AU semi-major axis, a 219±.2 day period, a .38±.01 ecc., and a 1.85M_J mini. mass.

Both planets cross the habitable zone enough to prevent stable orbits of terrestrial planets. The outer planet might have terrestrial-sized moons.

Orbital simulations show that out of 100 test planets, none are still in orbit after one million years.[79] Underwood, et al., report that a terrestrial planet orbit would not remain in the habitable zone for the last one billion years.[86]

HD 83443

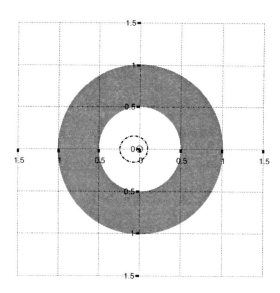

HD 83443 is a type K0V star in the constellation of Vela, The Sail, about 141.9 light-years away (RA 09 37 11.8281, D -43 16 19.939). It has a 8.2 apparent magnitude, a .79-.93$M_\odot$ mass, a 5,454±61°K effective temperature, and a .35±.08 FeH ratio. From its metallicity an age of 10.7 Gyrs is estimated, although others have calculated 3.2 Gyrs.

The inner planet, the fortieth discovered (2000, La Silla, European Southern Observatory, Chile[28]), with a .038AU semi-major axis (about 3.5 million miles), a 2.985 day period, a .13 eccentricity, and a .34±.01M_J minimum mass.

The outer planet, the forty-eighth discovered (2000, La Silla, European Southern Observatory, Chile[29]), has a .17AU semi-major axis, a 29.83 day period, a .42 eccentricity, and a .15M_J minimum mass (half the mass of Saturn), making it the smallest extra-solar planet so far. This planet was challenged, refined measurements confirmed it, more measurements denied it. It is still uncertain.

Both planets have no effect on the habitable zone. Depending on the correct age estimate, this planet could easily be in the equivalent of our dinosaur age or developed far beyond us.

Orbital simulations show that out of 100 test planets, eighty-three are still in orbit after one million years (simulations of our solar system yields only eighty-one out of 100).[79] Underwood, et al., report that a terrestrial planet orbit could remain in the habitable zone for the last one billion years.[86]

HD 108147

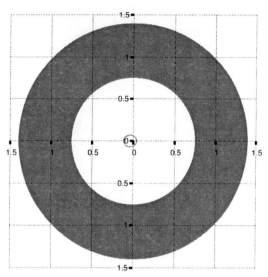

HD 108147 is a type F9 or G0V star in the constellation of Crux, The Cross, about 125.7 light-years away (RA 12 25 46.2686, D - 64 01 19.516). It has a 7.0 apparent magnitude, a $1.23M_\odot$ mass, a $6,248\pm42°K$ effective temperature, a $.20\pm.05$ FeH ratio, and a 8-day rotation. From its metallicity and rapid rotation, an age of 1–2 Gyrs is estimated.

The planet, the forty-first discovered (2000, La Silla, European Southern Observatory, Chile[28]), has a $.079\pm.021$AU semi-major axis, a $10.9\pm.001$ day period, a $.48\pm.09$ eccentricity, and a $.4\pm.01M_J$ estimated mass, $1.34M_S$, only one-third more than Saturn.

The planet is well below the inner edge of the habitable zone, having very little impact on any possible terrestrial-sized planets in the orbital zone.

There is the possibility that another planet is in the system, masked by the "jitter" of the rapid rotation of the chromosphere. If there is such a planet it has a 1.6AU semi-major axis, a $.42M_J$ mass and a 587 day period. If this planet exists then its gravity would prevent any stable orbits in the habitable zone. The discoverers of the inner planet are continuing their observation in an attempt to confirm or discredit this tentative conclusion.

Simulations show that out of 100 test planets, seventy-eight are still in orbit after one million years.[79] Underwood, et al., report that a terrestrial planet orbit could remain in the habitable zone for the last one billion years.[86]

HD 168746

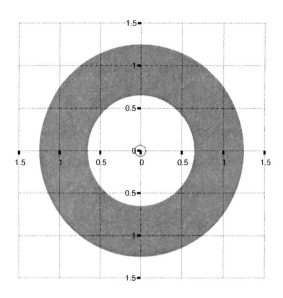

HD 168746 is a type G5 star in the constellation Scutum, The Shield, about 140.5 light-years away (RA 18 21 49.7832, D -11 55 21.660). It has an apparent magnitude of 7.5, mass of .88-.92$M_\odot$, a 5,601°K effective temperature, and a -.08±.05 FeH ratio. From its metallicity, just a little bit less than our Sun at -.07 (where our Sun equals 1), an age of 14.8 Gyrs is estimated. The chromosphere shows little to no activity supporting this age estimate.

The planet, the forty-second discovered (2000, La Silla, European Southern Observatory, Chile[28]), has a semi-major axis of .066±.01AU, a period of 6.400±.004 days, an eccentricity of .081±.081, and has an estimated mass of .24±.01M_J (77% that of Saturn). It has an approximate surface temperature of 900°K.

The planet is well below the habitable zone, having very little impact on any possible terrestrial-sized planets in the habitable zone.

Orbital simulations show that out of 100 test planets, 84 are still in orbit after one million years (simulations of our solar system yields only eighty-one out of 100).[79] Underwood, et al., report that a terrestrial planet orbit could remain in the habitable zone for the last one billion years.[86]

HD 169830

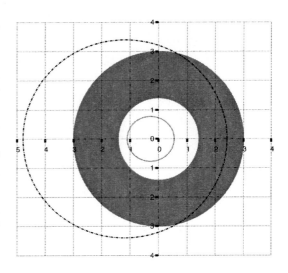

HD 169830 is a type F8V star in the constellation Sagittarius, The Archer, about 118.4 light-years away (RA 18 27 49.4838, D -29 49 00.715). It has an apparent magnitude of 5.8±.1, a mass of 1.41–1.43$M_\odot$, effective temperature is 6,299±41°K, and a .21±.05 FeH metallicity ratio. From its metallicity an age of 2–4 Gyrs is estimated. A rotation of 8.3–9.5 days seems to support this age estimate.

The planet, the forty-third discovered (2000, La Silla, European Southern Observatory, Chile[28]), has a semi-major axis of .823AU, a period of 225±.22 days, a .34±.05 eccentricity, and an estimated mass of 2.88M_J.

The outer planet, the 114th discovered (2003, La Silla, European Southern Observatory, Chile[55]), has a semi-major axis of 3.60AU, a period of 2,102±264 days, an eccentricity of .33±.02, and an estimated mass of 4.04M_J.

Underwood, et al., report that a terrestrial planet orbit could remain in the habitable zone for the last one billion years.[86]

Distance from Sun to:
Mercury = .387AU (period = 88 days)
Venus = .7233AU (period = 225 days)
Earth = 1AU (period = 365.242 days)
Mars = 1.52AU (period = 687 days)
Jupiter = 5.22AU (period = 3.64 years)
Saturn = 9.53AU (period 29.46 years)
Uranus = 19.19AU (period = 84.07 years)
Neptune = 30.6AU (period = 164.8 years)
Pluto = 39.5 AU (period = 248.53 years)

HD 38529

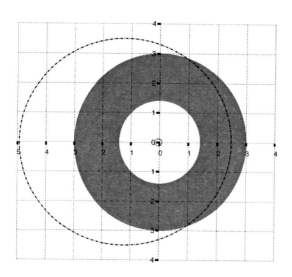

HD 38529 is a type G4IV star, a subgiant, in the constellation Orion, The Hunter, about 138 light-years away (RA 05 46 34.9120 D +01 10 05.496). It has a 5.95 apparent magnitude, a 1.39$M_\odot$ mass, a 2.82$R_\odot$ radius, a 5,674±40°K effective temperature, and a .40±.06 FeH ratio. The star is chromospherically quiet and appears to have a 34.5-day rotation rate.

The inner planet was the forty-fourth discovered (2000, Lick Observatory and Keck Observatory[25]), has a .129AU semi-major axis, a 14.31±.05 day period, a .28 eccentricity, and a .78M_J minimum mass.

The outer planet was the eighty-second discovered (2003 Lick Observatory, Keck Observatory[42]), has a 3.71±.03AU semi-major axis, a 2,207±33 day period, a .33 eccentricity, and a 12.78±.08 minimum mass. If the mass estimate for the primary is understated, then this object will be classified as a T brown dwarf.

Crossing into the habitable zone, the outer planet prevents stable orbits there. Any terrestrial moons would need to be very close to the outer planet to receive enough heat to compensate for the part of the orbit outside the habitable zone. Such close orbits would quickly tidally lock them, rendering them uninhabitable.

Orbital simulations show that out of 100 test planets, none are still in orbit after one million years.[79] It is more likely to support an asteroid belt than any planetoids Earth-moon sized or larger. Underwood, et al., report that a terrestrial planet orbit would not remain in the habitable zone for the last one billion years.[86]

HD 92788

HD 92788 is a type G5V star in the constellation Sextan, The sextant, about 106.9 light-years away (RA 10 42 48.5287, D -02 11 01.521). It has a 7.31 apparent magnitude, a $1.06M_\odot$ mass, a $5,821\pm41°K$ effective temperature, a $.32\pm.05$ FeH ratio, a $1.1L_\odot$ luminosity, and a $1.05R_\odot$ radius. The star is chromospherically quiet and appears to have a 31.7-day rotation. Its age is unknown.

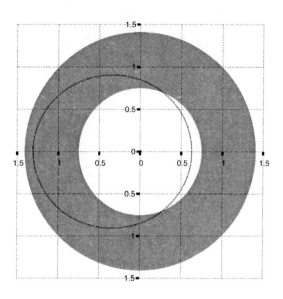

The planet (2000, Lick Observatory and Keck Observatory[25]) has a .96AU semi-major axis, a 325.0 day period, a $.35\pm.01$ eccentricity, and a $3.58M_J$ minimum mass.

Crossing into the habitable zone as it does, you would expect there to be no stable orbits for terrestrial planets. This is not so. If the terrestrial planet is in the right position, about .82AU from the primary and on the opposite side of the primary from the giant planet, it will have an orbital period of 300 days that will allow stability. The giant planet's orbit will move around the primary at the same rate as the terrestrial planet. In other words, from the surface of the terrestrial planet, the giant planet must be visible only in the early morning or evening, just as we see Venus, except it can never be seen to transit the primary. The giant planet can only be seen to appear and disappear behind the sun.

Orbital simulations show that out of 100 test planets, none are still in orbit after one million years.[79] Underwood, et al., report that a terrestrial planet orbit would not remain in the habitable zone for the last one billion years.[86]

HD 6434

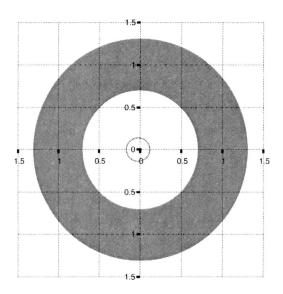

HD 6434 is a type G3IV subgiant star in the constellation Phoenix, about 131.4 light-years away (RA 01 04 40.1511, D -39 29 17.583). It has a 7.72 apparent magnitude, a .79$M_\odot$ mass, a 1.12$L_\odot$ luminosity, a 5,835±50°K effective temperature, a -.52±.05 FeH ratio, and a 18.5-day rotation. Another source has the mass listed as .80$M_\odot$ with an age of at least 17 Gyrs, although 3.7 Gyrs is the generally accepted age. This star has the lowest metallicity of any star hosting planets so far detected.

Given that the star's mass is equal to or less than our Sun and that the luminosity is higher, this star appears to be moving through the subgiant stage, especially if the age of 17 Gyrs is correct (which would make it a .8$M_\odot$ mass).

The planet, the forty-sixth found (2000, La Silla, European Southern Observatory[29]), is well below the inner edge of the habitable zone, with a .14AU semi-major axis, a 21.99 day period, an .17±.03 eccentricity, and a .39M_J minimum mass.

Given the planet's close orbit, the habitable zone is available for terrestrial planets in stable orbits. However, given the star's apparent age, only planets orbiting at the outer edge to middle of the zone would still be habitable.

Orbital simulations show that out of 100 test planets, fifty-eight are still in orbit after one million years.[79] Underwood, et al., report that a terrestrial planet orbit could remain in the habitable zone for the last one billion years.[86]

HD 19994 (94 Ceti)

HD 19994 (94 Ceti) is a type F8V star in the constellation Cetus, The Whale, about 73 light-years away (RA 03 12 46.4365, D -01 11 45.964). It has a 5.07 apparent magnitude, a 1.35$M_\odot$ mass, a 3.87$L_\odot$ luminosity, a 6,190±57°K effective temperature, a .24±.07 FeH ratio, and a 13.7-day rotation. It has a 3.1 Gyrs age estimate. It also has a close M dwarf companion (GJ 128B) about 100AU away.

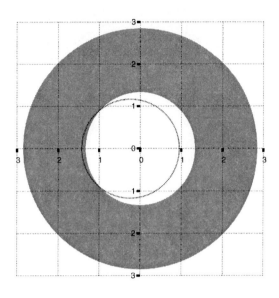

The planet, the forty-seventh discovered (2000, La Silla, European Southern Observatory[29]), has a 1.19±.11AU semi-major axis, a 454±19 day period, a .2 eccentricity, and a 1.66±.34M_j minimum mass.

The planet just barely intrudes into the habitable zone, but given its large gravitational influence it's more than enough to make it difficult for any planets in the zone. No stable orbits can survive within about 1–1.5 AU of the giant planet, meaning any planets have to exist at the outer edges of the habitable zone, making them rather cool places to live. Orbital simulations show that out of 100 test planets, one is still in orbit after one million years.[79] Underwood, et al., report that a terrestrial planet orbit could remain in only part of the habitable zone for the last one billion years.[86]

HD 121504

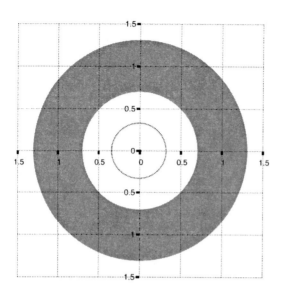

HD 121504 is a type G2V star in the constellation Centaurus, The Centaur, about 144.6 light-years away (RA 13 57 17.2375, D -56 02 24.153). It has a 7.54 apparent magnitude, a 1.11–1.17$M_\odot$ mass, a 1.58$L_\odot$ luminosity, a 6,075±40°K temperature, a .16±.07 FeH ratio, and a 14.8-day rotation. Its age is estimated at between 2.8 and 4 Gyrs, with the fast rotation leaning towards the younger age.

The planet, the forty-ninth discovered (2000, La Silla, European Southern Observatory[29]), has a .33AU semi-major axis, a 63.33 day period, a .03±.01 eccentricity, and a 1.22M_J minimum mass.

The planet is close to the primary and thus has only limited gravitational influence over the habitable zone. However, stable orbits will still tend to be from the middle of the zone outward.

Solar System Object	Apparent Magnitude
Sun	-26.5
Full Moon -	12.5
First Quarter Moon	-10.20
Last Quarter Moon	-10.05
Venus	-4.6
Jupiter, Mars	-2.9
Naked eye limit	6.5

Orbital simulations show that out of 100 test planets, sixty are still in orbit after one million years (simulations of our solar system yields only eighty-one out of 100).[79] Underwood, et al., report that a terrestrial planet orbit could remain in the habitable zone for the last one billion years.[86]

HD 190228

HD 190228 is a type G5IV star in the constellation Vulpecula, The Fox, about 215 light-years away (RA 20 03 00.7730, D +28 18 24.685). It has a 7.3 apparent magnitude, a $1.23\pm.19M_\odot$ mass (although the discoverers of this planet put the star's mass at .84M), a 5,327°K temperature, and a -.26 FeH ratio. It has a 4.5 Gyrs estimated age, which would be consis-

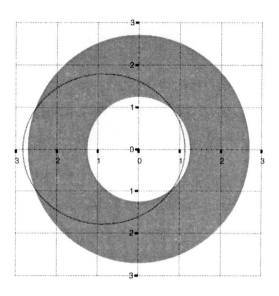

tent with its subgiant status and a $1.2M_\odot$ mass. This star has a low metallicity, among the lowest of the stars with discovered planets. One paper has suggested that one explanation for this low metallicity is that the planet is instead a Brown dwarf with a 33 to $53M_J$ mass.

The planet, the fiftieth discovered (2000, La Silla, European Southern Observatory[29]), has a 1.98±.33AU semi-major axis, a 1,112±42 day period, a .43±.08 eccentricity, and a 3.44–$5.0M_J$ minimum mass. One team, using optical equipment claims the object has a $45M_J$ mass and a 6.3° orbital inclination.

Passing through the habitable zone as it does the planet prevents stable orbits. However, skirting both the inner and outer edges of the zone as it does lends to the possibility of habitable terrestrial-sized moons. The summers would be scorching (heat like the Sahara desert) and winters truly awful (cold like Antarctica).

Orbital simulations show that out of 100 test planets, none are still in orbit after one million years.[79] Underwood, et al., report that a terrestrial planet orbit would not remain in the habitable zone for the last one billion years.[86]

HD 22049 (Epsilon Eridani)

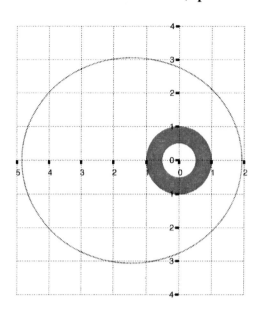

HD 22049 (Epsilon Eridani) is a type K2V star. It is the 18th brightest star in the constellation Eridanus, The River, about 10.43 light years away (RA 03 32 55.8442, D -09 27 29.744). It has a 3.73 apparent magnitude, a $.85 M_\odot$ mass, a $.292 L_\odot$ luminosity, a $.86 D_\odot$ diameter, a 5,073±42°K temperature, and a -.13±.04 FeH ratio. It has a .7–2 Gys age estimate. It has an active chromosphere, supporting this young age conclusion. At about 60AU there is also a dust ring.

The planet, the fifty-first discovered (2000, McDonald Observatory, Lick Observatory, La Silla European Southern Observatory, and Canada-France-Hawaii Telescope[30]), has a 3.39±.07AU semi-major axis, a 2,550±48 day period, a .43 eccentricity, and a .88 M_J minimum mass.

The dust ring appears to have a gap that could be explained by the presence of a planet at 40AUs, a period of greater than 140 years, and a .1M_J mass. The eccentricity would be .3 or more.

The presence of a planet outside the habitable zone tends to have a greater influence on the stability of orbits than a planet that is below the zone. Thus, while the giant planet only approaches to within 1AU of the habitable zone, that's more than enough to disrupt any planets orbiting in the zone.

Orbital simulations show that out of 100 test planets, none are still in orbit after one million years.[79] Underwood, et al., report that a terrestrial planet orbit would not remain in the habitable zone for the last one billion years.[86]

HD 179949

HD 179949 is a type F8V star in the constellation Sagittarius, The Archer, and is about 88 light-years away (RA 19 15 33.2278, D -24 10 45.668). It has a 6.25 apparent magnitude, a 1.24–1.28$M_\odot$ mass, a 1.24$D_\odot$ diameter, a 6,260±43°K effective temperature, a .22±.05 FeH ratio, and a moderate 9-day rotation.

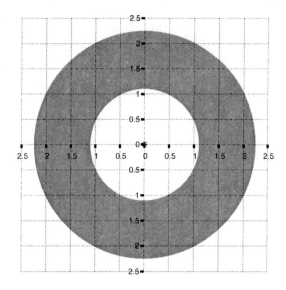

The planet (2000, Anglo-Australian Telescope and Keck Observatory[31]) has a .045±.005AU semi-major axis, a 3.092±.001 day period, a 0.0±.03 eccentricity, and a .93±.05M_J minimum mass.

The planet is far enough from the habitable zone to render its gravitational influence moot. Because an F-type star is brighter and hotter than our own, it burns through its fuel faster. With the listed mass, the star will spend about 3.5 Gyrs on the main sequence. Whether there will be sufficient time for life to develop fully on a terrestrial planet is questionable, but not impossible.

Orbital simulations show that out of 100 test planets, eighty-eight are still in orbit after one million years (simulations of our solar system yields only eighty-one out of 100).[79] Underwood, et al., report that a terrestrial planet orbit could remain in the habitable zone for the last one billion years.[86]

HD 27442 (Epsilon Reticuli)

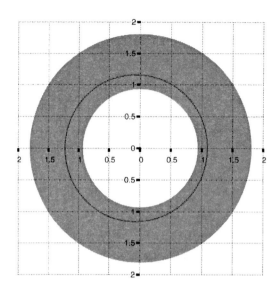

HD 27442 (Epsilon Reticuli) is a type K2IVa star, the fifth brightest star in the constellation Reticulum, The Net, about 59 light-years away (RA 04 16 29.0287, D -59 18 07.763). It has a 4.55 apparent magnitude, a 1.2M$_\odot$ mass, a 4.86L$_\odot$ luminosity, a 4,825K temperature, a .39±.13 FeH ratio, a 4.5D$_\odot$ diameter, and a 10 Gyrs estimated age.

The planet, the fifty-third discovered (2000, Anglo-Australian Telescope[32]), has a 1.16±.02AU semi-major axis, a 415±146 days period, a .06±.04 eccentricity, and a 1.32±.11M$_J$ minimum mass.

Although the planet is inside the habitable zone, there is still the possibility of stable terrestrial planetary orbits at the outer edge of the zone. This is because of the nearly circular orbit of the planet, which allows a "resonance" to develop between the planets. This resonance is a simple ratio, such as 1:2 or 2:3. That is for every three orbits of the giant planet, the terrestrial planet orbits twice. This is a very stable configuration as long as there are no other giant planets in the system that are closer than about 4–5AU. The down side is that it places the terrestrial planet at the outside edge of the habitable zone, although as the star gets older, and hotter, the middle of the zone will move out to the planet and warm it up.

Terrestrial-sized moons around the giant planet would be habitable, given the right atmosphere; perhaps even several could be if they could establish a "resonant" orbital situation.

Orbital simulations show that out of 100 test planets, eight are still in orbit after one million years.[79] Underwood, et al., report that a terrestrial planet orbit would only remain in the outer portion of the habitable zone for the last one billion years.[86]

HD 160691 (Mu Arae, Gliese 691)

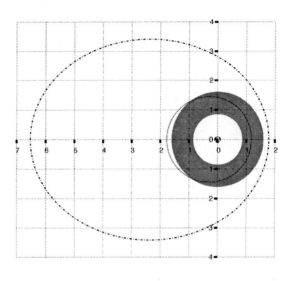

HD 160691 (Mu Arae, Gliese 691) is a type G3IV-V star, although Hipparcos suggests it is a type G5V. It is the twelfth brightest star in the constellation Ara, The Altar, and it is about 49.87 light-years away (RA 17 44 08.7029, D - 51 50 02.591). It has a 5.20 apparent magnitude, a 1.08–1.10$M_\odot$ mass, a 1.81$L_\odot$ luminosity, a 5,798°K effective temperature, a .263±.03 FeH ratio, a 1.43$D_\odot$ diameter, and a 4.5–6 Gyrs estimated age (although it could be as young as 2Gyrs). It is chromospherically inactive, photometrically stable, and seems to have a rotational period of 31 days, which supports the 4.5Gyr age estimate.

The planet, the fifty-fourth discovered (2000, Anglo-Australian Telescope[32]), has a 1.48±.1AU semi-major axis, a 638±10 day period, a .2±.08 eccentricity, and a 1.68±.2M_J minimum mass.

The other planet, the 125th found (2004, La Silla, European Southern Observatory, Chile[63]), is inside the other planet's orbit. It has a .09AU semi-major axis, a 9.55±.03 day period, a 0.0±.02 eccentricity, and a minimum mass of fourteen times that of the Earth (14M_E). Using an albedo of .35 the planet's temperature is about 900°K.

The outermost planet, the 135th found (2004, Anglo-Australian Telescope[69]), has a 4.16AU semi-major axis, an 8.2 year period, a .57 eccentricity, and a 3.1M_J minimum mass. Only 70% of its orbit has been observed.

The middle planet orbits through the habitable zone, but spends a significant amount of time just beyond the outer edge of the zone. Heat from the planet might be enough to allow for a habitable terrestrial-sized moon.

Orbital simulations show that out of 100 test planets, none are still in orbit after one million years.[79] Underwood, et al., report that a terrestrial planet orbit would not remain in the habitable zone for the last one billion years.[86]

HD 8574

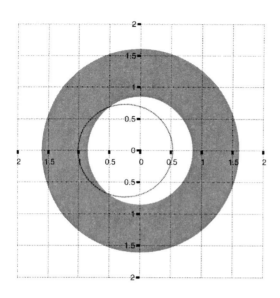

HD 8574 is a type F8 star in the constellation Pisces, The Fishes, about 143.9 light-years away (RA 01 25 12.5168, D +28 34 00.096). It has a 7.12 apparent magnitude, a 1.17$M_\odot$ mass, a 2.25$L_\odot$ luminosity, a 6,151°K temperature, and a .06±.07 FeH ratio.

The planet, the fifty-sixth discovered (2001, Haute-Provence Observatory, France[34]), has a .77±.01AU semi-major axis, a 228.18±.68 day period, a .31±.09 eccentricity, and a 2.08±.17M_J min. mass.

The habitable zone is dominated by the gravity of the giant planet, thus there are no stable orbits in the zone for terrestrial-sized planets. Because the planet goes in as close as .5AU, terrestrial-sized moons would be uninhabitable.

Orbital simulations show that out of 100 test planets, none are still in orbit after one million years.[79] Underwood, et al., report that a terrestrial planet orbit could be confined only to the outer portion of the habitable zone for the last one billion years.[86]

HD 28185

HD 28185 is a type G5V star in the constellation Eridanus, about 128.4 light-years away (RA 04 26 26.3205, D -10 33 02.955). It has a 7.8 apparent magnitude, a .99$M_\odot$ mass, a 1.02$L_\odot$ luminosity, a 5,656°K temperature, and a .22±.05 FeH ratio. It has a 2.4 and 4 Gyrs estimated age.

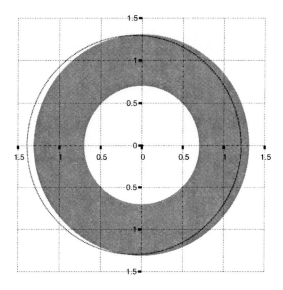

The planet, the fifty-seventh found (2001, La Silla, European Southern Observatory, Chile[34]), has a 1.03AU semi-major axis, a 383±2 day period, a .07±.04 eccentricity, and a 5.7M_J estimated minimum mass.

Skirting the outside edge of the habitable zone as it does, you would expect there to be no stable orbits for terrestrial-sized planets in the zone. This isn't true as simulations have shown. The nearly circular orbit would allow another planet to establish a stable orbit in "resonance" with the giant planet. The resonance most likely is a 1:2 arrangement, which would place the terrestrial planet at the inner edge of the habitable zone, with a year of about 200 days.

Orbital simulations show that out of 100 test planets, three are still in orbit after one million years.[79] Underwood, et al., report that a terrestrial planet orbit could remain in the inner portion of the habitable zone for at least one billion years.[86]

HD 50554

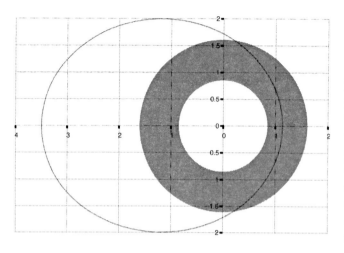

HD 50554 is a type F8V star in the constellation Gemini, The Twins, about 101.15 light-years away (RA 06 54 42.8253, D +24 14 44.011). It has a 6.84 apparent magnitude, a 1.1±.04$M_\odot$ mass, a 1.096$R_\odot$ radius, a 6,026±30°K effective temperature, a .01±.04 FeH ratio, and a 16.1-day rotation. It has a 3.9–4.5 Gyrs age estimate.

The planet was the fifty-eighth discovered (2001, Haute-Provence Observatory, France[34]) and it has a 2.32±.06AU semi-major axis, a 1,254±25 day period, a .51±.01 eccentricity, and a 3.72±1.8M_J minimum mass.

Cutting through the habitable zone the way it does completely disrupts any possible orbits inside the zone. As far as it gets from its primary, any terrestrial moons would be uninhabitable.

Orbital simulations show that out of 100 test planets, none are still in orbit after one million years.[79] Underwood, et al., report that a terrestrial planet orbit would not remain in the habitable zone for the last one billion years.[86]

HD 74156

HD 74156 is a type G0 star in the constellation Hydra, The Water Serpent, about 210.4 light-years away (RA 08 42 25.1222, D +04 34 41.151). It has a 7.61 apparent magnitude, a $1.05\pm.01M_\odot$ mass, a $3.12L_\odot$ luminosity, a 6,112°K effective temperature, and a $.16\pm.05$ FeH ratio. Its age is unknown.

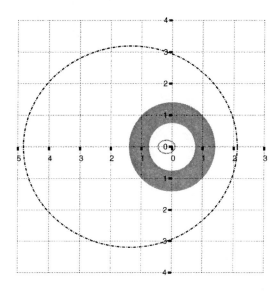

The inner planet, the fifty-ninth discovered (2001, La Silla, European Southern Observatory, Chile[34]), has a .276AU semi-major axis, a $51.60\pm.053$ day period, a $.65\pm.022$ eccentricity, and a $1.55\pm.01M_J$ minimum mass.

The outer planet, the sixtieth discovered (2001, La Silla, European Southern Observatory, Chile[34]), has a 3.47AU semi-major axis, a 2,300.0 day period, a $.39\pm.074$ eccentricity, and a $7.46\pm.04M_J$ minimum mass.

The inner planet's orbit brings it close enough to the habitable zone for its gravity to disrupt any orbits in the zone. The influence of the more massive outer planet makes it impossible to find a stable orbital pattern in the zone.

Orbital simulations show that out of 100 terrestrial test planets, none are still in orbit after one million years. One Moon-sized particle survived for 100Myrs.[79] Underwood, et al., report that a terrestrial planet orbit would not remain in the habitable zone for the last one billion years.[86]

HD 80606

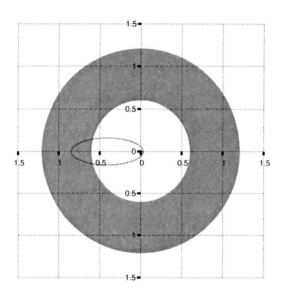

HD 80606 is a type G5 star in the constellation Ursa Major, The Larger Bear, about 190.3 light-years away (RA 09 22 37.5679, D +50 36 13.397). It is a visual binary, that is, the companion star, HD 80607, is visible through a telescope, and the separation distance is about 1,800AU. The distance is far enough that any conclusions about its orbit regarding period and eccentricity cannot be made, although its orbital period is estimated at over 1 Gyr.

HD 80606 has a 9.06 apparent magnitude, a .9±.13$M_\odot$ mass, a .75$L_\odot$ luminosity, a 5,574±72°K effective temperature, and a .32±.09 FeH ratio.

HD 80607's is also a type G5 star, with a 9.86 apparent magnitude, a 1.15$M_\odot$ mass, a .95$L_\odot$ luminosity, and a 5,555°K effective temperature.

The planet, the sixty-first discovered (2001, Keck Observatory and Haute-Provence Observatory, France[34]), has a .438±.001 semi-major axis, a 111.7±.21 day period, a .93±.012 eccentricity, and a 3.43±.02M_J minimum mass. Its eccentricity is unusually high and the only models that explain this high eccentricity require another planet with a period of about 100 years. None has so far been detected.

The habitable zone is dominated by the intrusion of the highly eccentric orbit of the giant planet, making impossible stable orbits in that zone.

Orbital simulations show that out of 100 test planets, none are still in orbit after one million years.[79] Underwood, et al., report that a terrestrial planet orbit would not remain in the habitable zone for the last one billion years.[86]

HD 106252

HD 106252 is a type G0 star in the constellation Virgo, The Virgin, about 122 light-years away (RA 12 13 29.5093, D +10 02 29.898). It has a 7.36 apparent magnitude, a $1.05\pm.03M_\odot$ mass, a $1.27L_\odot$ luminosity, a $1.096R_\odot$ radius, a 5,899°K effective temperature, a $-.01\pm.05$ FeH ratio, and a 22.8-day rotation. It has a 5 Gyr estimated age.

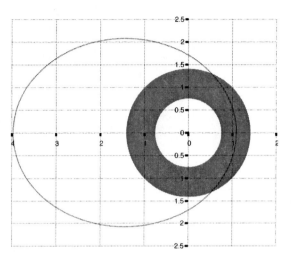

The planet, the sixty-third discovered (2001, La Silla, European Southern Observatory, Chile[34]), has a $2.53\pm.01$AU semi-major axis, a $1,503\pm1.5$ day period, a .57 eccentricity, and a $6.79\pm.02M_J$ minimum mass.

Because the planet cuts through the habitable zone, there are no stable orbits possible in it. Ranging as far from its primary as the planet does, there is no possibility for habitable moons orbiting it.

Orbital simulations show that out of 100 test planets, none are still in orbit after one million years.[79] Underwood, et al., report that a terrestrial planet orbit would not remain in the habitable zone for the last one billion years.[86]

HD 141937

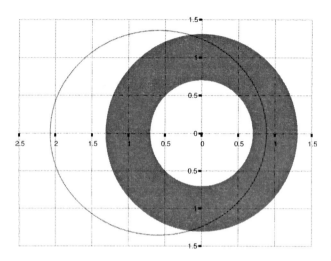

HD 141937 is a type G2/3V star in the constellation Libra, The Scales, about 109 light-years away (RA 15 52 17.5474, D -18 26 09.834). It has a 7.25 apparent magnitude, a $1.08-1.1 M_\odot$ mass, a $1.17 L_\odot$ luminosity, a 5,909°K effective temperature, a .10±.05 FeH ratio, and a 13.25-day rotation. Its age estimate is 1.6–6 Gyrs, with 2.5 Gyrs being the most likely.

The planet, the sixty-fourth discovered (2001, La Silla, European Southern Observatory, Chile[34]), has a 1.48±.04AU semi-major axis, a 658±6 day period, a .4±.01 eccentricity, and a 9.67±.03M_J minimum mass.

The habitable zone is dominated by the presence of the giant planet, making impossible stable orbits inside the zone. Habitable terrestrial moons of the giant planet are not likely given the time spent outside the habitable zone.

Orbital simulations show that out of 100 test planets, none are still in orbit after one million years.[79] Underwood, et al., report that a terrestrial planet orbit would not remain in the habitable zone for the last one billion years.[86]

HD 178911B

HD 178911B is a type G5 star in the constellation Lyra, about 152.3 light-years away (RA 19 09 03.1039, D +34 35 59.454). It is a member of a triple binary system, where HD 178911A and 178911B are a visual binary with a separation of 640±240AU, and HD 178911Aa and HD 178911Ab are a spectroscopic binary with a 2.3AU semi-major axis, a .59 eccentricity, and a 3.55 year period.

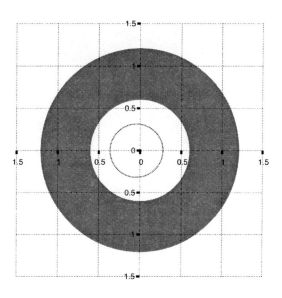

HD 178911B has a 7.97 apparent magnitude, .87-.98$M_\odot$ a mass, a 1.3$L_\odot$ luminosity, a 5,600±42°K effective temperature, and a .27±.05 FeH ratio.

The planet, the sixty-fifth discovered (2001, Keck Observatory and Haute-Provence Observatory, France[34]), has a .32±.06AU semi-major axis, a 71.50±.013 day period, a .14±.016 eccentricity, and a 6.46±.17M_J minimum mass.

The habitable zone is relatively free of gravitational influences, but the further from the inner edge the terrestrial planet is the more likely it will have a stable orbit.

The 178911Aa-b binary pair orbit too far apart from each other for stable planetary orbits in their habitable zones.

Orbital simulations show that out of 100 test planets, sixty-three are still in orbit after one million years (simulations of our solar system yields only eighty-one out of 100).[79] Underwood, et al., report that a terrestrial planet orbit could remain in the habitable zone for the last one billion years.[86]

HD 213240

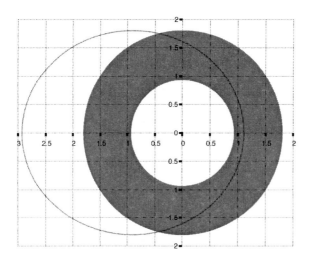

HD 213240 is a type G4IV star in the constellation Grus, The Crane, about 131.8 light-years away (RA 22 31 00.3672, D -49 25 59.773). It has a 6.81 apparent magnitude, a 1.22$M_\odot$ mass, a 2.59$L_\odot$ luminosity, a 5,984°K effective temperature, a .17±.05 FeH ratio, and a 15-day rotation. It has a 2.7–2.9 Gyr age estimate.

The planet, the sixty-sixth discovered (2001, La Silla, European Southern Observatory, Chile[34]), has a 2.02AU semi-major axis, a 951 day period, a .45 eccentricity, and a 4.49M_J min. mass.

The habitable zone is dominated by the giant planet that passes through it, preventing stable orbits for terrestrial planets in the zone. Terrestrial-sized moons of the giant planet would not be habitable because of the time spent outside the habitable zone.

Orbital simulations show that out of 100 test planets, none are still in orbit after one million years.[79] Underwood, et al., report that a terrestrial planet orbit would not remain in the habitable zone for the last one billion years.[86]

HD 4203

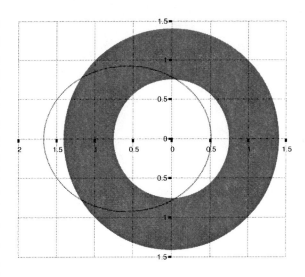

HD 4203 is a type G5V star in the constellation Pisces, The Fishes, about 252 light-years away (RA 00 44 41.2021, D +20 26 56.138). It has a 8.68 apparent magnitude, a 1.06±.13M$_\odot$ mass, a 5,636°K effective temperature, a .40±.05 FeH ratio, and is chromospherically quiet. Others have a 98M$_\odot$ estimated stellar mass, but that estimate doesn't take into account the high metallicity of the star. It has a 7.6 Gyr age estimate, although others have calculated a 8.7 Gyr estimate.

The planet, the sixty-eighth discovered (2001, Lick Observatory and Keck Observatory[36]), has a 1.09AU semi-major axis, a 406.0±3 day period, a .53±.02 eccentricity, and a 1.64M$_J$ minimum mass.

The habitable zone is dominated by the presence of the giant planet, there is no chance that stable orbits exist in the zone.

Orbital simulations show that out of 100 test planets, none are still in orbit after one million years.[79] In contradiction to other studies, Underwood, et al., report that a terrestrial planet orbit could remain in part of the habitable zone for the last one billion years.[86]

HD 4208

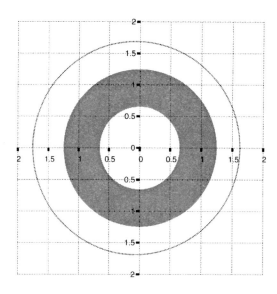

HD 4208 is a type G5V star in the constellation Sculptor, The Sculptor's Tools, about 110.5 light-years away (RA 00 44 26.6503, D -26 30 56.449). It has a 7.79 apparent magnitude, a $.93\pm.07 M_\odot$ mass, a 5,626°K effective temperature, a -24±.04 FeH ratio, and an age estimated at 11.7 Gyrs (standard measurements place it at 4.3 Gyrs). It is rotating slowly and is chromospherically inactive, supporting the older age estimate.

The planet, the sixty-ninth discovered (2001, Lick Observatory and Keck Observatory[36]), has a 1.69AU semi-major axis, a 829.0±2 day period, a .04 eccentricity, and a $.81 M_J$ min. mass.

The regular orbit of the giant planet and its distance from the habitable zone means that there are many possibilities for stable terrestrial planetary orbits. The further the terrestrial planet's orbit is from the giant planet, the more stable its own orbit.

Orbital simulations show that out of 100 test planets, sixty-seven are still in orbit after one million years.[79] The Lagrange Points, L_4 and L_5, also show great stability, with .1 eccentricity, or less, keeping the planet in the Habitable Zone. The instability zone starts at 1.3AU and any eccentricity of .2 or greater. Also, resonant orbits at .8, .84, .86, and .87AU are unstable.[80] Underwood, et al., report that a terrestrial planet orbit would only remain in the inner portion of the habitable zone for the last one billion years.[86]

HD 33636

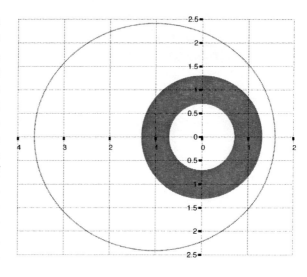

HD 33636 is a type G0V star in the constellation Orion, The Hunter, about 93.5 light-years away (RA 05 11 46.4490, D +04 24 12.742). It has an apparent magnitude of 7.06, a mass of $.99\pm.13M_\odot$, an effective temperature 6,046°K, a -.08±.064 FeH ratio, and an estimated age of 2.5–2.8 Gyrs.

The planet, the seventieth discovered (2001, Lick Observatory and Keck Observatory[36]), has a 2.62±.2AU semi-major axis, a 1,553±67 day period, a .39±.02 eccentricity, and a 7.71±.11M_J minimum mass. Because the time of observation for this planet candidate was shorter than its projected orbital period, there is a large uncertainty in the number.

The eccentricity of the giant planet's orbit, and the fact that it approaches to with .25AU of the habitable zone ruins the chances for stable planetary orbits in the zone. The large mass of the giant planet means its gravity has a strong influence over any object in the habitable zone. Close encounters would highly disturb the orbit of the smaller planet, eventually ejecting it from the system or crashing it into the star.

Orbital simulations show that out of 100 test planets, none are still in orbit after one million years.[79] Underwood, et al., report that a terrestrial planet orbit would not remain in the habitable zone for the last one billion years.[86]

HD 68988

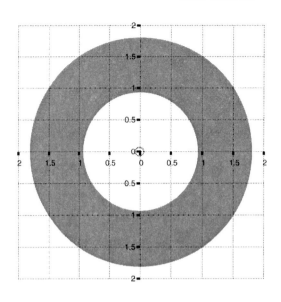

HD 68988 is a type G0 star in the constellation Ursa Major, The Larger Bear, about 189 light-years away (RA 08 18 22.1731, D +61 27 38.599). It has a 8.21 apparent magnitude, a 1.2$M_\odot$ mass, a 5,988±52°K effective temperature, and a .36±.06 FeH ratio. Its age is estimated age at 1.6 Gyrs (based on metallicity) or 6–7.1 Gyrs (based on chromospheric inactivity and its slow rotation).

The planet, the seventy-first discovered (2001, Lick Observatory and Keck Observatory[36]), has a .07±.001AU semi-major axis, a 6.276 day period, a .14 eccentricity, and a 1.9M_J estimated mass. Based on its close orbit, the eccentricity is too high. There are three explanations: 1) the tidal drag estimates are too low; 2) the planet has recently (in the last 100 million years) arrived at its orbit; or 3) its being perturbed by a third giant body in the system. Supporting this last possibility is a linear trend in the orbital speed of the planet that could be explained by the presence of a third body with a period much greater than four years.

Being so close to its primary and so far from the habitable zone the giant planet has no affect on the stability of orbits in the habitable zone. This assumes that if there is a second giant planet in the system it is far enough from the habitable zone to keep its influence to a minimum. A giant planet orbiting slightly further from its primary than Jupiter is from our sun (5.22AU) would have the four year period and little impact on terrestrial planet orbits in the habitable zone.

Orbital simulations, based only on one giant planet, show that out of 100 test planets, eighty-four are still in orbit after one million years (simulations of our solar system yields only eighty-one out of 100).[79] Underwood, et al., report that a terrestrial planet orbit could remain in the habitable zone for the last one billion years.[86]

HD 114783

HD 114783 is a type K2V star in the constellation Virgo, The Virgin, about 71.72 light-years away (RA 13 12 43.7860, D -02 15 54.143). It has a 7.57 apparent magnitude, a .92±.04$M_\odot$ mass, a .34$L_\odot$ luminosity, a 5,098°K effective temperature, and a -.09±.04 FeH ratio. It has a 4.8–6.0 Gyr age estimate. The star is chromospherically inactive.

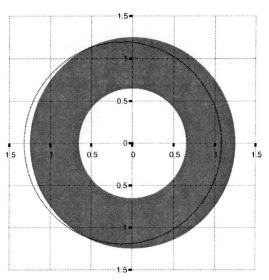

The planet, the seventy-second discovered (2001, Lick Observatory and Keck Observatory[36]), has a 1.2AU semi-major axis, a 501±1.14 day period, a .1±.01 eccentricity, and a .99±.02M_J minimum mass.

Orbiting at the outer edge of the habitable zone, the giant planet dominates the zone, severely restricting the possibility of stable orbits for smaller planets.

Terrestrial-sized moons around the planet, however, would be habitable. The extra heat contributed to the moons by the giant planet would compensate for the distance from the primary. The moons wouldn't be hot, but neither would they be freezing cold.

Orbital simulations show that out of 100 test planets, one is still in orbit after one million years.[79] Underwood, et al., report that a terrestrial planet orbit could be confined only to the inner portion of the habitable zone for the last one billion years.[86] A study of the Lagrange Points indicates a stable area extending from 1.16AU to 1.24AU, and between 30° and 110° for L_4 (leading the Gas Giant).

HD 142A

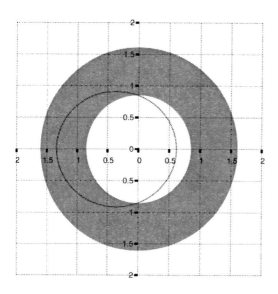

HD 142A is a type G1IV star in the constellation Phoenix, about 83.5 light-years away (RA 00 06 19.1755, D -49 04 30.688). It is the dominant member of a binary system with a separation of 138AU. It has a 5.71 apparent magnitude, a $1.1\pm.16M_\odot$ mass, a $2.98L_\odot$ luminosity, a $1.61D_\odot$ diameter, a 6,302°K effective temperature, a $.14\pm.07$ FeH ratio, and it is chromospherically inactive.

The second star, HD 142B, is a type MV star, with an 11.5 absolute magnitude, a $.34M_\odot$ mass, a $.014L_\odot$ luminosity, and a $.11D_\odot$ diameter.

From A, B would appear as a bright -11.41 star, brighter than a quarter-moon appears from Earth. From B, A would appear as a -17.2 star, almost a hundred times brighter than a full moon appears on Earth.

The planet, the seventy-third discovered (2001, Anglo-Australian Telescope[37]), has a .980AU semi-major axis, a 331.8±7.1 day period, a .37±.01 eccentricity, and a $1.14\pm.14M_J$ minimum mass.

The binary nature of the HD 412 system is probably a contributing factor to the eccentricity of the giant planet's orbit, which curves through the habitable zone. This high eccentricity prevents stable orbits in the habitable zone for other planets.

Orbital simulations show that out of 100 test planets, none are still in orbit after one million years.[79] Underwood, et al., report that a terrestrial planet orbit could remain in only part of the habitable zone for at least one billion years.[86]

HD 23079

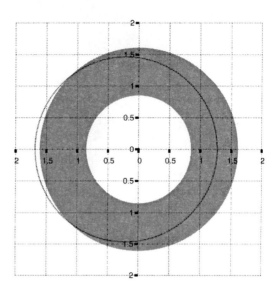

HD 23079 is a type F8/G0V (intermediate between F8 and G0) star in the constellation Horologium, The Clock, about 113.4 light-years away (RA 03 39 43.0952, D -52 54 57.017). It has a 6.47 apparent magnitude, a 1.1–1.25$M_\odot$ mass, and a 1.4$L_\odot$ luminosity, a .846$D_\odot$ diameter, a 5,959°K effective temperature, and a -.11±.06 FeH ratio. It appears to be very stable, that is, over the life of the Hipparcos satellite mission its light output did not change to the limits of the mission instruments.

The planet, the seventy-fourth discovered (2001, Anglo-Australian Telescope[37]), has a 1.48±.01AU semi-major axis, a 628±1 day period, a .14±.1 eccentricity, and a 2.76±.23M_J minimum mass.

With the giant planet orbiting inside the habitable zone, there is no possibility of stable orbits for other planets in the zone. Terrestrial-sized moons, though, should do well, as the extra heat from the giant planet will stave off the worst of the cold caused by being at the outer edge of the zone.

Orbital simulations show that out of 100 test planets, none are still in orbit after one million years.[79] Underwood, et al., report that a terrestrial planet orbit could remain in only part of the habitable zone for at least one billion years.[86] However, a study of the Lagrange Points, L_4 and L_5, indicate planets at these locations would have very stable orbits with 0.1 eccentricity or less.

HD 39091 (Pi Mensae)

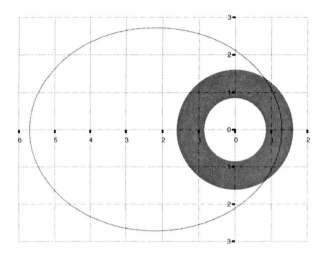

HD 39091 (Pi Mensae) is a type GIV star in the constellation Mensa, The Table Mountain, and is about 102.69 light-years away (RA 05 37 09.8917, D -80 28 08.839). It is the 16th brightest star in that constellation with a 5.65 apparent magnitude, a 1.1$M_\odot$ mass, a 1.557$L_\odot$ luminosity, a 5,991°K effective temperature, a .10±.04 FeH ratio, and is chromospherically inactive.

The planet, the seventy-fifth discovered (2001, Anglo-Australian Telescope[38]), has a 3.5±.21AU semi-major axis of, a 2,280±212 day period, a .63±.01 eccentricity, and a 10.39±.04M_J minimum mass.

Passing through the habitable zone as it does the giant planet would disrupt any stable orbits for other planets.

Orbital simulations show that out of 100 test planets, none are still in orbit after one million years.[79] Underwood, et al., report that a terrestrial planet orbit could remain in only the inner portion of the habitable zone for at least one billion years.[86]

HD 137759 (Iota Draconis)

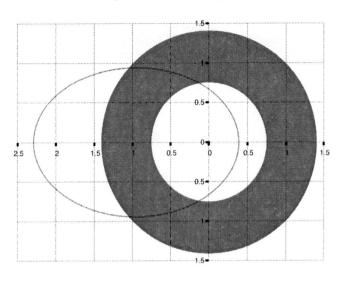

HD 137759 (Iota Draconis, HIP 75458) is a K2III star in the constellation Draco and is about 102.69 light-years away (RA 15 24 55.7747, D +58 57 57.836). It is the ninth brightest star in that constellation and has been listed as a wide separation (7,900AU) binary system (BD +59° 1655) since 1879. Modern measurements, however, show a velocity difference (which has not yet been verified) between the two stars greater than can be explained by a very eccentric orbit. Thus, the case for them being a binary system is weak. HD 137759 (HIP 75458) has a 3.3 apparent magnitude, a $1.05M_\odot$ mass, a $70L_\odot$ luminosity, a 4,775°K effective temperature, a .13±.14 FeH ratio, and a $13D_\odot$ diameter. It is also listed as a variable star.

The planet, the seventy-sixth discovered (2001, Lick Observatory[39]), has a 1.34AU semi-major axis, a 550.0±.6 day period, a .71 eccentricity, and a $8.68±.04M_J$ minimum mass. Hipparcos data fail to reveal any sign of a stellar companion, putting a $45M_J$ maximum limit to the object.

Passing through the Habitable zone as it does the giant planet would disrupt orbits for other planets in the zone.

Orbital simulations show that out of 100 test planets, none remain in orbit after one million years.[79] Underwood, et al., report that a terrestrial planet orbit would not remain in the habitable zone for the last one billion years, nor would a terrestrial planet orbit remain within the previous habitable zone when the star was on the main sequence portion of its development.[86]

HD 136118

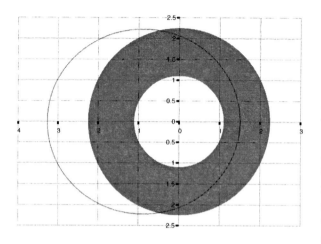

HD 136118 is a type F9V in the constellation Serpens Caput, The Serpent's Tail, about 170.5 light-years away (RA 15 18 55.4719, D -01 35 32.590). It has a 6.9 apparent magnitude (its absolute magnitude is 3.61, indicating that this star is already beginning to evolve off the main sequence), a $1.24\pm.24M_\odot$ mass, a $1.58D_\odot$ diameter, a 6,222°K effective temperature, a -.04±.05 FeH ratio, and has a chromosphere that is moderately active. It has a 12.2-day rotation and has a 3.2–3.6 Gyrs age estimate.

The planet, the seventy-seventh discovered (2002, Lick Observatory and Keck Observatory[40]), has a 2.39±.09AU semi-major axis, a 1,209±24 day period, a .37±.025 eccentricity, and a $11.91\pm.2M_J$ minimum mass. This is close to the limit between planet and brown dwarf so if the stellar mass estimate or orbital inclination are off then this object will be a brown dwarf. However, in this case, the measurements indicate that we are viewing the star along its equator. If the orbit is in the equatorial plane, then the minimum mass is the absolute mass.

In either case, the habitable zone has no stable orbits for other planets. If it is a planet, then terrestrial-sized moons would be uninhabitable. While a brown dwarf by itself would have only uninhabitable planets due to tidal-locking, the combined output of the primary and the brown dwarf might provide enough energy for terrestrial-sized planets to be habitable even though they spend significant amounts of time outside the primary star's habitable zone. The larger the brown dwarf, the more likely this scenario might become.

Orbital simulations show that out of 100 test planets, none are still in orbit after one million years.[79] Underwood, et al., report that a terrestrial planet orbit would not remain in the habitable zone for the last one billion years.[86]

HD 49674

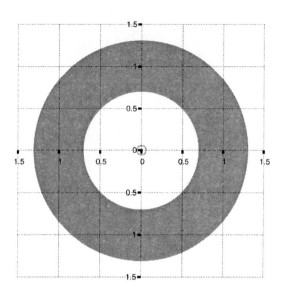

HD 49674 is a type G0V star, although the discoverers of the planet assigned it a type G5V which is more consistent with its blue/visual magnitude and a 5,600°K temperature. It is in the constellation Auriga, The Charioteer, about 132.5 light-years away (RA 06 51 30.5164, D +40 52 03.923). It has an 8.1 apparent magnitude, a mass equal to our Sun, a .815$L_\odot$ luminosity, a 5,644°K effective temperature, a .33±.06 FeH ratio, and a .982$D_\odot$ diameter.

The planet, the eighty-third discovered (2002, Keck Observatory[43]), has a .057±.003AU semi-major axis, a 4.948±.001 day period, a .00±.18 eccentricity (near zero), and a .12M_J minimum mass.

Located far below the inner boundary of the habitable zone, the giant planet has no effect on the stability of orbits of any planets located in the zone.

Orbital simulations show that out of 100 test planets, eighty-one are still in orbit after one million years (simulations of our solar system yields only eighty-one out of 100).[79] Underwood, et al., report that a terrestrial planet orbit could remain in the habitable zone for the last one billion years.[86]

HD 72659

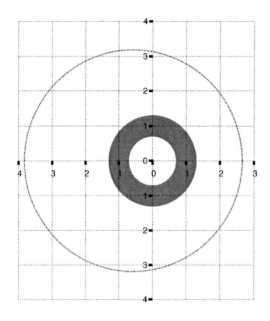

HD 72659 is a type G0V star in the constellation Hydra, The Water Serpent, about 167.5 light-years away (RA 08 34 03.1895, D -01 34 05.583). It has a 7.8 apparent magnitude, a .95M$_\odot$ mass, a 5.995±45°K effective temperature, and a .03±.06 FeH ratio, and is chromospherically quiet.

The planet, the eighty-fourth discovered (2002, Keck Observatory[43]), has a 3.24AU semi-major axis, a 2,185 day period, a .18 eccentricity, and a 2.54±.01M$_J$ minimum mass. The period is uncertain because the time spent observing has so far been less than one orbit.

The giant planet is far enough outside the habitable zone that it has minimal effect on the orbits of planets in the zone.

Orbital simulations show that out of 100 test planets, sixty are still in orbit after one million years (simulations of our solar system yields only eighty-one out of 100).[79] The major constraint is that the planet have low eccentricity, .2 or below.[80] Underwood, et al., report that a terrestrial planet orbit could remain in the habitable zone for the last one billion years.[86] This system is one of the few in which the giant planet stays outside the habitable zone for the entire main sequence life of the star, much like the giant planets in our own solar system.

HD 108874

HD 108874 is a type G5V star in the constellation Coma Berenices, Berenice's Hair, about 223 light-years away (RA 12 30 26.8829, D +22 52 47.383). It has a 8.76 apparent magnitude, and a $1.0\pm.04M_\odot$ mass, a $1.25L_\odot$ luminosity, a 5,596°K effective temperature, and a $.23\pm.05$ FeH ratio. Currently the dimmest G-dwarf star observed with Doppler velocities, it has about a 23% higher metallicity than our sun.

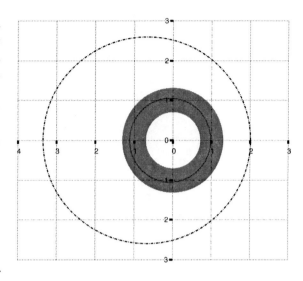

The inner planet, the eighty-fifth discovered (2002, Keck Observatory[43]), has a $1.05\pm.02$AU semi-major axis, a 395.4 ± 2.5 day period, a $.07\pm.04$ eccentricity, and a $1.36\pm.13M_J$ minimum mass.

Presently, the planet orbits inside the habitable zone, making it impossible for any others to be there. Terrestrial-sized moons would be habitable.

The outer planet, the 155th discovered (2005, Keck Observatory[99]), has a $2.68\pm.25$AU semi-major axis, a $1,605.8\pm88$ day period, a $.25\pm.07$ eccentricity, and a $1.018\pm.3M_J$ minimum mass.

Orbital simulations show that out of 100 test planets, none are still in orbit after one million years.[79] Underwood, et al., report that a terrestrial planet orbit would only remain in part of the habitable zone for at least one billion years, when the habitable zone was sufficiently far from the giant planet for stable orbits.[86]

HD 114729

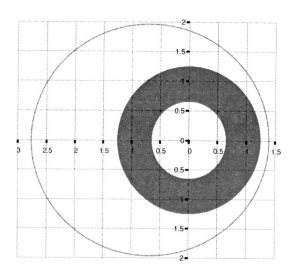

HD 114729 is a type G0V star in the constellation Centaurus, The Centaur, about 114 light-years away (RA 13 12 44.2575, D -31 52 24.056). SIMBAD has it listed as a G3V, but Hipparcos and the discoverers of the planet derived G0V based on its temperature, metallicity, and difference between absolute and apparent magnitudes. It has a 6.69 apparent magnitude, a .93±.01$M_\odot$ mass, a 5,886°K effective temperature, and a -.25±.05 FeH ratio. The star is chromospherically quiet and has a 3.95 absolute magnitude, which implies the star has evolved half a magnitude above the zero-age main sequence. The chromosphere indicates an age of 5 Gyrs.

The planet, the eighty-sixth discovered (2002, Keck Observatory[43]), has a 2.08AU semi-major axis, a 1,136±2.5 day period, a .33±.01 eccentricity, and a .88±.02M_J min. mass.

Coming so close to the outer edge of the habitable zone, the giant planet would disrupt the orbits of any planets in the zone to point of making it impossible for stable orbits to exist. Orbiting as far from the zone as it does means any terrestrial-sized moons would also be uninhabitable.

Orbital simulations show that out of 100 test planets, none are still in orbit after one million years.[79] Underwood, et al., report that a terrestrial planet orbit would only remain in the inner portion of the habitable zone for the last one billion years.[86]

HD 128311

HD 128311 is a type K3V star, although SIMBAD lists it as a K0V, in the constellation Bootes, The Herdsman, about 54.1 light-years away (RA 14 36 00.5607, D +09 44 47.466). It has a 7.51 apparent magnitude, a $.8M_\odot$ mass, a 4,835°K effective temperature, and a .03±.07 FeH ratio. The star is young, and chromospherically active, making a difficult case for the planet candidate.

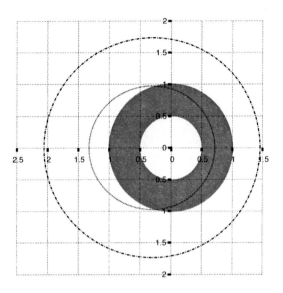

The inner planet, the eighty-seventh announced (2002, Keck Observatory[43]), has a 1.02AU semi-major axis, a 420.514 day period, a .3 eccentricity, and a $2.58M_J$ minimum mass.

The giant planet's gravity dominates the habitable zone preventing stable orbits for other planets. A planet at the inner edge (~.5AU) might be habitable, as might terrestrial-sized moons of the giant planet.

The outer planet, the 156th announced (2005, Keck Observatory[99]), has a 1.76±.13AU semi-major axis, a 919 day period, a .17±.09 eccentricity, and a $3.21\pm.3M_J$ minimum mass.

Underwood, et al., report that a terrestrial planet orbit would not remain in the habitable zone for the last one billion years.[86]

HD 30177

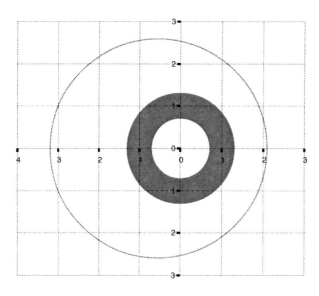

HD 30177 is a type G8V star in the constellation Reticulum, The Net, about 179 light-years away (RA 04 41 54.3731, D -58 01 14.725). It has an 8.41 apparent magnitude, a .95±.05$M_\odot$ mass, a 5,584±65°K effective temperature, and a .38±.09 FeH ratio. It is chromospherically inactive and appears to beginning to evolve off the main sequence.

The planet, the eighty-eighth discovered (2002, Anglo-Australian Telescope[44]), has a 2.65±.01AU semi-major axis, a 1,620±2 day period, a .21±.17 eccentricity, and a 7.64±.05M_J minimum mass. Although the eccentricity is listed as .21, the error range, ±.17, implies that the planet could actually have a circular orbit with no eccentricity.

The giant planet's gravity has a significant effect on orbits in the habitable zone, although it is no closer than about .9AU to the zone. Safe orbital zones are probably closer to the inside border of the zone than the outer border.

Orbital simulations show that out of 100 test planets, ten are still in orbit after one million years.[79] Underwood, et al., report that a terrestrial planet orbit would only remain in part of the habitable zone for the last one billion years.[86]

HD 73526

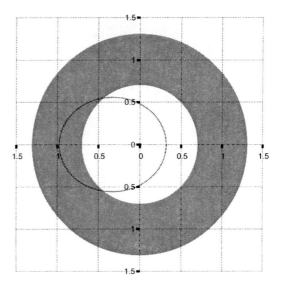

HD 73526 is a type G6V star in the constellation Vela, The Sails (Of The Ship), about 322.7 light-years away (RA 08 37 16.4839, D -41 19 08.767). It has a 9.0 apparent magnitude, a $1.02M_\odot$ mass, a 5,699±49°K effective temperature, and a .27±.06 FeH ratio.

The planet, the eighty-ninth discovered (2002, Anglo-Australian Telescope[44]), has a .647±.013AU semi-major axis, a 188.0±2 day period, a .52±.21 eccentricity, and a 3.63±.75M_J minimum mass.

The giant planet intrudes into the habitable zone for part of its orbit, preventing stable orbits in the zone.

Orbiting as close as it does to the primary, any terrestrial-sized moons would be uninhabitable.

Orbital simulations show that out of 100 test planets, one is still in orbit after one million years.[79] Underwood, et al., report that a terrestrial planet orbit would only remain in part of the habitable zone for the last one billion years.[86]

HD 196050

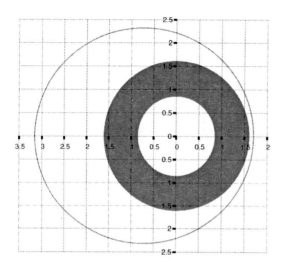

HD 196050 is a type G3V star in the constellation Pavo, The Peacock, about 152.9 light-years away (RA 20 37 51.7102, D -60 38 04.135). It has a 7.6 apparent magnitude, a 1.1–1.15$M_\odot$ mass, a 5,918°K effective temperature, and a .22±.05 FeH ratio.

The planet, the ninetieth discovered (2002, Anglo-Australian Telescope[45]), has a 2.43AU semi-major axis, a 1,321±54 day period, a .3 eccentricity, and a 3.02M_j minimum mass.

The giant planet's eccentric orbit brings its gravitational influence to cover most of the habitable zone, reducing significantly the probability that terrestrial-sized planets could have stable orbits in the zone.

However, if there were a 1:3 orbital resonance, a planet at 1.17AU (429 day period) could have a stable orbit. A 1:4 resonance yields a .88AU orbit (322 days). How such planets could form from a dust disk with the interference of the giant planet is a different matter.

Orbital simulations show that out of 100 test planets, one is still in orbit after one million years.[79] Underwood, et al., report that a terrestrial planet orbit would only remain in the inner portion of the habitable zone for the last one billion years.[86]

HD 216437

HD 216437 is a type G4IV/V star is in the constellation Indus, The Indian, about 86.4 light-years away (RA 22 54 39.4833, D -70 04 25.352). It has a 6.06 apparent magnitude, a 1.15–1.20M$_\odot$ mass, a 5,887°K effective temperature, and a .25±.04 FeH ratio. It is chromospherically inactive.

The planet, the ninety-first discovered (2002, Anglo-Australian Telescope[45]),

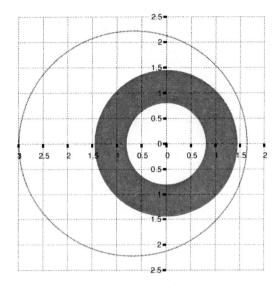

has a 2.32AU semi-major axis, a 1,256±35 day period, a .29±.12 eccentricity, and a 1.82M$_J$ minimum mass.

The giant planet's eccentric orbit brings its gravitational influence to cover most of the habitable zone, reducing significantly the probability of terrestrial-sized planets with stable orbits in the zone.

However, if there were a 1:3 orbital resonance, a planet at 1.18AU (431 day period) could have a stable orbit. A 1:4 resonance yields a .88AU orbit (323 days). How such planets could form from a dust disk with the interference of the giant planet is a different matter.

Orbital simulations show that out of 100 test planets, two are still in orbit after one million years.[79] Underwood, et al., report that a terrestrial planet orbit would only remain in the inner portion of the habitable zone for the last one billion years.[86]

HD 20367

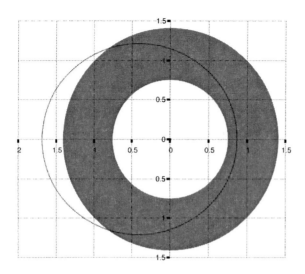

HD 20367 is a type G0 star is in the constellation Aries, The Ram, about 88 light-years away (RA 03 17 40.0461, D +31 07 37.372). It has a 6.4 apparent magnitude, a $1.17M_\odot$ mass of, a $1.67L_\odot$ luminosity, a 6,138°K effective temperature, and a .17±.10 FeH ratio.

The planet, the ninety-second discovered (2002, Haute-Provence Observatory and Geneva Observatory[46]), has a 1.28±.03AU semi-major axis, a 500±9.3 day period, a .32±.09 eccentricity, and a $1.12±.05M_J$ minimum mass.

Spending most of its orbit inside the habitable zone, the giant planet prevents any other planets from having stable orbits there. Terrestrial-sized moons would have interesting seasons as they would have great temperature extremes. The giant planet would provide some heat to keep the planet warm, plus the reflection of light from its atmosphere would perhaps help as well.

Orbital simulations show that out of 100 test planets, none are still in orbit after one million years.[79] Underwood, et al., report that a terrestrial planet orbit would only remain in part of the habitable zone for at least one billion years.[86]

HD 23596

HD 23596 is a type F8 star is in the constellation Perseus, about 169.5 light-years away (RA 03 48 00.3739, D +40 31 50.287). It has a 7.24 apparent magnitude, a 1.29±.01M$_\odot$ mass, a 2.76L$_\odot$ luminosity, a 6,108±36°K effective temperature, and a .31±.05 FeH ratio.

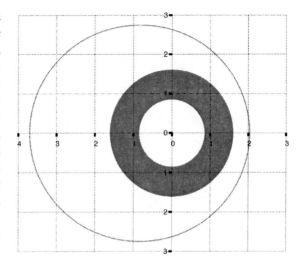

The planet, the ninety-third discovered (2002, Haute-Provence Observatory and Geneva Observatory[46]), has a 2.87±.01AU semi-major axis, a 1,548±25 day period, a .298±.028 eccentricity, and a 8.0±.81M$_J$ minimum mass.

Coming to within .5AU of the habitable zone, the giant planet's gravity dominates the zone and prevents stable orbits for other planets in the zone. Being so far from the primary, this planet's terrestrial-sized moons would be uninhabitable.

Orbital simulations show that out of 100 test planets, none are still in orbit after one million years.[79] Underwood, et al., report that a terrestrial planet orbit would not remain in the habitable zone for the last one billion years.[86]

HD 150706

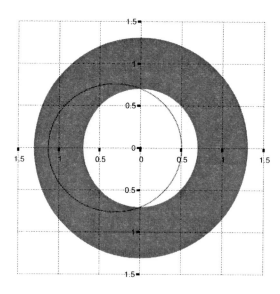

HD 150706 is a type G0 star in the constellation Ursa Minor, The Smaller Bear, about 88.6 light-years away (RA 16 31 17.5856, D +79 47 23.189). It has a 7.02 apparent magnitude, a .98±.23$M_\odot$ mass, a .98$L_\odot$ luminosity, a 5,961°K effective temperature, and a -.01±.04 FeH ratio.

The planet, the ninety-fourth discovered (2002, Haute-Provence Observatory and Geneva Observatory[46]), has a .82AU semi-major axis, a 264.9±5.8 day period, a .38±.12 eccentricity, and a minimum mass equal to Jupiter.

Cutting through the habitable zone as it does, this giant planet prevents other planets in the zone from having stable orbits. Terrestrial moons would have scorching hot summers and cool winters.

Orbital simulations show that out of 100 test planets, none are still in orbit after one million years.[79] Underwood, et al., report that a terrestrial planet orbit could remain in the habitable zone for the last one billion years.[86]

HD 190360 (Gliese 777A)

HD 190360 (Gliese 777A) is a type G6IV star in the constellation Cygnus, The Swan, about 51.8 light-years away (RA 20 03 37.4055, D +29 53 48.500). It has a 5.7 apparent magnitude of, a $.90\pm.06M_\odot$ mass, a $1.18L_\odot$ luminosity, a $5,584\pm36°K$ effective temperature, and a $.24\pm.05$ FeH ratio.

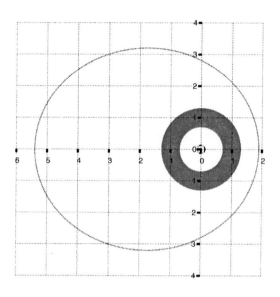

This is apparently a triple star system, although one measurement places the other two stars at a distance of almost 4 light-years from Gl777A. Gl 777Ba is a type M4.5–6 V, a dim red dwarf star. The star has a .19M mass, a .19D diameter, and a .0004 luminosity. It appears to have a faint companion star Gl 777Bb with a 69.9 AU separation.

Gl 777 Bb may be a type M4.5–6 V dim red dwarf star. There are no firm numbers regarding mass, size, or luminosity.

The outer planet, the ninety-fifth discovered (2002, Haute-Provence Observatory and Geneva Observatory[46]), has a $3.92\pm.2$AU semi-major axis, a $2,891\pm85$ day period, a .36 eccentricity, and a $1.502\pm.13M_J$ minimum mass.

The inner planet, the 157th discovered (2005, Keck Observatory[99]), has a $.128\pm.002$AU semi-major axis, a $17.1\pm.015$ day period, a .01 eccentricity, and a $.057\pm.015M_J$ minimum mass.

Orbital simulations, based on the outer planet's presence only, show that out of 100 terrestrial test planets, eighty-six are still in orbit after one million years (simulations of our solar system yields only eighty-one out of 100).[79] Another group's simulation shows that nearly the entire Habitable Zone is stable.[80] Underwood, et al., report that a terrestrial planet orbit could remain in the habitable zone for the last one billion years.[86] The inner planet's close orbit and low mass shouldn't affect these results.

HD 114386

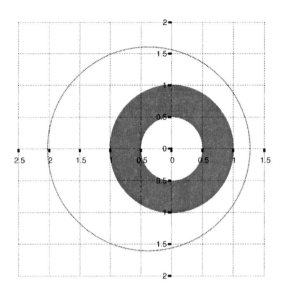

HD 114386 is a type K3V star in the constellation Centaurus, The Centaur, about 91.28 light-years away (RA 13 10 39.8231, D -35 03 17.218). It has an 8.8 apparent magnitude, a .54-.75$M_\odot$ mass, a .28$L_\odot$ luminosity, a 4,804°K effective temperature, and a -.08±.06 FeH ratio.

The planet, the ninety-sixth discovered (2002, Haute-Provence Observatory and Geneva Observatory[46]), has a 1.65AU semi-major axis, a 937±15.6 day period, a .23 eccentricity, and a 1.24M_J minimum mass.

The habitable zone is dominated by the gravity of the giant planet, which severely restricts the probability of stable orbits. However, there is the possibility for "resonance" between the giant planet and a planet in the zone. An example would be a 1:3 resonance, a planet at .79AU with a period of 290 days. For every one orbit of the giant planet, the terrestrial planet would have three orbits. The most stable arrangement would have the terrestrial planet on the opposite side of the star from the giant planet when the giant planet is closest to the habitable zone.

Orbital simulations show that out of 100 test planets, eleven are still in orbit after one million years.[79] Underwood, et al., report that a terrestrial planet orbit could remain in the habitable zone for the last one billion years.[86] They add that this is one of the few systems like our own where the giant planets do not intrude into the habitable zone for the entire main sequence life of the star.

HD 147513

HD 147513 is a type G3/5V star in the constellation Scorpius, The Scorpion, about 42 light-years away (RA 16 24 01.2899, D -39 11 34.729). It has a 5.3 apparent magnitude, a .92±.19$M_\odot$ mass, a luminosity equal to our Sun, a 5,883°K effective temperature, a .06±.04 FeH ratio, a 4.7-day rotation, and an estimated age of 300 million years.

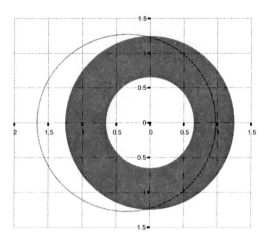

HD 147513 is a proposed barium dwarf with a common proper motion with a white dwarf at a distance of 5,360AU. Researchers explain the barium origin by the process of mass transfer in a binary system in which the secondary gained matter from the evolved primary, now a white dwarf. The large separation is explained by the ejection of the white dwarf from a quadruple system including HD 147513. It is worth noting that the activity level, age, and metallicity support membership in the Ursa Major kinematic group, the proposed origin of the white dwarf and HD 147513.

The planet, the ninety-seventh discovered (2002, Haute-Provence Observatory and Geneva Observatory[46]), has a 1.32AU semi-major axis, a 528±6.3 day period, a .26±.05 eccentricity, and a 1.21M_J minimum mass.

Cutting through the habitable zone as it does, the giant planet makes it impossible for stable orbits to exist in the zone for other planets. And with its large eccentricity, terrestrial-sized moons would have very difficult seasons.

Because of the age of this system, it is possible the "moons" would not yet have solid surfaces, and the atmosphere's would be primordial "soups" of gases. Orbital simulations show that out of 100 test planets, none are still in orbit after one million years.[79] Underwood, et al., report that a terrestrial planet orbit would not remain in the habitable zone for at least billion years.[86]

HD 2039

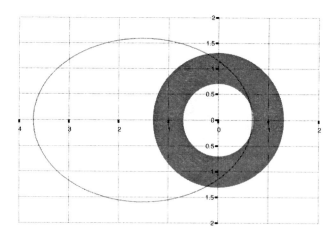

HD 2039 is a type G2/3IV/V star in the constellation Phoenix, about 292.7 light-years away (RA 00 24 20.2778, D -56 39 00.171). It has a 9.0 apparent magnitude, a .98± .22M$_\odot$ mass, a 5,976°K effective temperature, a .32±.06 FeH ratio, and is chromospherically inactive.

The planet, the ninety-eighth discovered (2002, Anglo-Australian Telescope[44]), has a 2.2±.02AU semi-major axis, a 1,190.0±7 day period, a .69 eccentricity, and a 5.1±.27M$_J$ minimum mass.

The giant planet cuts completely through the habitable zone, rendering it impossible for stable orbits inside the zone.

Underwood, et al., report that a terrestrial planet orbit would not remain in the habitable zone for the last one billion years.[86]

M$_\odot$	=	Mass of our Sun (1.989 × 10^{30} kilograms)
L$_\odot$	=	Luminosity of our Sun, 3.90 × 10^{26} watts
R$_\odot$	=	Radius of our Sun = 695,000 km, 434,000 miles
D$_\odot$	=	Diameter of our Sun = 1,390,000 km, 868,000 miles
M$_J$	=	Jupiter's Mass = .000955M$_\odot$
D$_J$	=	Jupiter's diameter = 142,984 km, 88,789 miles
M$_S$	=	Saturn's Mass = .000285M$_\odot$, .298M$_J$
M$_N$	=	Neptune's Mass = .0005128M$_\odot$ = .0536M$_J$ = .184M$_S$
M$_E$	=	Earth's Mass = .000003M$_\odot$ = .003146M$_J$, = .0105MS = .0581M$_N$
AU	=	149.6 million km, 92,960,116 million miles
K	=	Kelvin, the absolute temperature scale. 273.15°K = 0°Centigrade, 32°F (water freezes), 373.15°K = 100°K, 212°F (water boils)

HD 76700

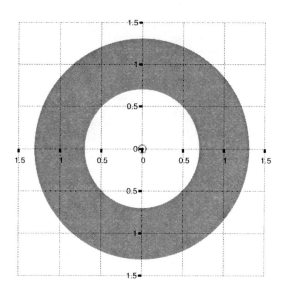

HD 76700 is a type G6V star (although different sources list it as G5 or G8V) in the constellation Volans, The Flying Fish, about 194.6 light-years away (RA 08 53 55.5153, D -66 48 03.571). It has a 8.13 apparent magnitude, a 1.0±.05$M_\odot$ mass, a 5,737±34°K effective temperature, and a .41±.05 FeH ratio.

The planet, the ninety-ninth discovered (2002, Anglo-Australian Telescope[44]), has a .049±.004AU semi-major axis, a 3.971±.001 day period, a 0.0±.04 eccentricity (nearly zero), and a .19±.017M_J minimum mass (about .63 times the mass of Saturn).

The giant planet is far enough below the lower edge of the habitable zone that it exerts little to no influence over the stability of any planets orbiting in the zone.

Underwood, et al., report that a terrestrial planet orbit could remain in the habitable zone for the last one billion years.[86]

HD 216435 (Tau Gruis)

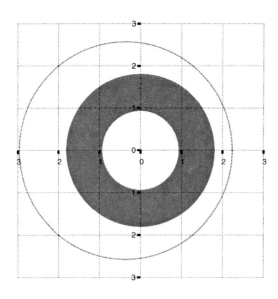

HD 216435 (Tau Gruis) is a type G0V (Hipparcos assigns it a type of G3IV) star in the constellation Grus, The Crane, about 108.5 light-years away (RA 22 53 37.9315, D -48 35 53.828). It has a 6.03 apparent magnitude, a 1.25–1.34$M_\odot$ mass, a 5,938°K effective temperature, and a .24±.05 FeH ratio. The apparent magnitude at the calculated distance puts this star a full magnitude above the main sequence and explains the difference in the assigned spectral types. The star is chromospherically inactive and with the other data gives an implied age of 5 Gyrs and implies the star has evolved off the main sequence.

The planet, the 100th discovered (2002, Anglo-Australian Telescope[47]), has a 2.6AU semi-major axis, a 1,326 day period, a .14 eccentricity, and a 1.23M_J minimum mass.

The habitable zone is dominated by the gravity of the giant planet, which severely restricts the probability of stable orbits. However, there is the possibility for "resonance" between the giant planet and a planet in the zone. An example would be a 1:2 resonance, a planet at 1.81AU with a period of 663 days. For every one orbit of the giant planet, the terrestrial planet would have two orbits. The most stable arrangement would have the terrestrial planet on the opposite side of the star from the giant planet when the giant planet is closest to the habitable zone. Or a 1:3 resonance, a planet at 1.21AU with a period of 442 days. For every one orbit of the giant planet, the terrestrial planet would have two orbits.

Underwood, et al., report that a terrestrial planet orbit would not remain in the habitable zone for the last one billion years.[86]

HD 222404 (Gamma Cephi, Elrai)

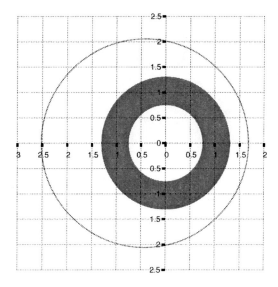

HD 222404 (Gamma Cephi, Elrai) is a type K1IV star in the constellation Cephi, Cepheus, about 38.4 light-years away (RA 23 39 20.8490, D +77 37 56.193). It has a 3.22 apparent magnitude, a 1.6$M_\odot$ mass, a 4.7$D_\odot$ diameter, a 4,916°K effective temperature, and a .16±.08 FeH ratio. Its age is estimated at 3 Gyrs.

This star is a binary, where the second star is a type M1V, with a .4$M_\odot$ mass, a .5$D_\odot$ diameter, and a 3,500°K effective temperature. It orbits the primary at a 12–32AU distance, with a 74 year period, and has a .44 eccentricity.

The planet, the 101st discovered (2002, McDonald Observatory[48]), has a 2.1AU semi-major axis, a 903±6 day period, a .209 eccentricity, and a 1.76M_J minimum mass.

The orbital simulations for this star indicate that there is a stable orbit at 1–1.15 AU from the primary, up to an inclination of 10°. Stable orbits at .8AU and below require a 15° to 40° inclination.[81] Underwood, et al., report that a terrestrial planet orbit would only remain in the inner portion of the habitable zone for the last one billion years.[86]

HD 40979

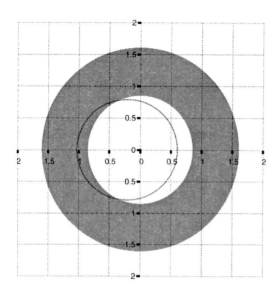

HD 40979 is a type F8V star in the constellation Auriga, The Charioteer, about 108.5 light-years away (RA 06 04 29.9431, D +44 15 37.599). It has a 6.74 apparent magnitude, a 1.08$M_\odot$ mass, a 1.26$D_\odot$ diameter, a .21±.05 FeH ratio, and a 6,145±42°K effective temperature. The combination of mass and radius suggest the star is a slightly evolved G0IV star. However, taking other data into consideration gives a 1.5 Gyr age estimate.

The giant planet, the 102nd discovered (2002, Lick Observatory, Keck Observatory[42]), has a .818±.07AU semi-major axis, a 260±7 day period, a .26±.03 eccentricity, and a 3.16±.16M_J min. mass.

The habitable zone is dominated by the giant planet's gravity, probably preventing stable orbits for terrestrial-sized planets.

If an orbital "resonance" of 2:1 is used, it might be possible for a planet at 1.42AU with a period 520 days to have a stable orbit, putting it right at the outer edge of the habitable zone.

Orbital simulations show that out of 100 test planets, none are still in orbit after one million years.[79] Underwood, et al., report that a terrestrial planet orbit would only remain in the outer portion of the habitable zone for the last one billion years.[86]

OGLE-TR-56

OGLE-TR-56 is a type G2V star in the constellation Sagittarius, The Archer, and is located 4,890 light-years away (RA 17 56 35.51 D -29 32 21.2). In fact, it is not even in our arm of the galaxy, the Orion Arm, but is in the Sagittarius Arm. It is very similar to our Sun, with a 1.04M$_\odot$ mass, a 1.1R$_\odot$ radius, a 5,900°K effective temperature, and a 4 Gyr age estimate. No apparent magnitude is given.

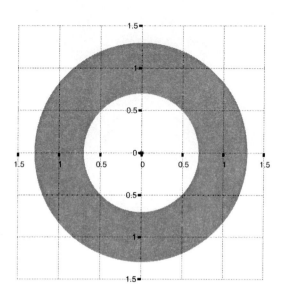

It is the 103rd planet found (2002, Whipple Observatory, AZ, and Las Campanas Observatory, Chile[49]), with a .0225AU semi-major axis (2 million miles), a 29 hour period, a zero eccentricity, and a 1.45±.23M$_J$ mass. It is incredibly hot at 1,686°K, has a 1.23±.16R$_J$ radius, and a density of .5 g/cm^3, about the same as Saturn's .69 g/cm^3. Being so close to its primary, the planet is probably face-locked with one side always facing its primary.

It is postulated that the planet was much larger in the past and had spiraled in even closer to the star. But as it got closer it began to lose mass, halting its spiral and reversing it as it lost mass until it reached its present position. In spite of its close orbit and high temperature, the planet is stable, although it is losing its atmospheric gases as they are excited by its proximity to the primary, much like HD 209458. In spite of the mass loss, the planet will most likely easily survive until its primary reaches nova stage.

The habitable zone is unaffected by the presence of the giant planet and the system is old enough to have not only inhabitable planets, but inhabited ones at that.

Underwood, et al., report that a terrestrial planet orbit could remain in the habitable zone for the last one billion years.[86]

HD 3651 (54 Piscis)

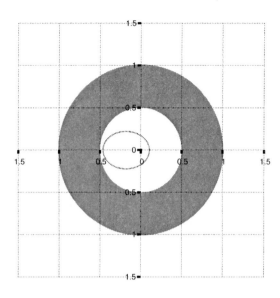

HD 3651 (54 Piscis) is a type K0V star in the constellation Pisces, The Fishes, about 36.2 light-years away (RA 00 39 21.8061, D +21 15 01.701). It has an apparent magnitude of 5.8, a mass of $.79M_\odot$, a luminosity of $0.479L_\odot$, a 5,173°K effective temperature, a .12±.04 FeH ratio, and a diameter of $1.015D_\odot$. It is also a variable star, that is, its light output varies over time.

The planet, the 104th discovered (2003, Lick Observatory, Keck Observatory[50]), has a .284AU semi-major-axis, a 62.23 day period, a .63 eccentricity, and a $.20M_J$ minimum mass.

The inner half of the habitable zone is probably dominated by the giant planet, but the outer regions might have stable orbits possible.

Underwood, et al., report that a terrestrial planet orbit could only be confined to the outer portion of the habitable zone for at least one billion years.[86]

HD 47536

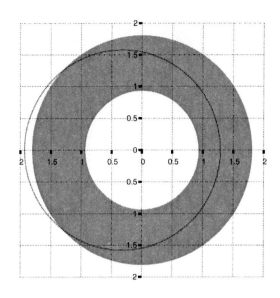

HD 47536 is a type K0III star in the constellation Pisces, The Fishes, about 395 light-years away (RA 06 37 47.6189, D -32 20 23.045). It has a 5.2 apparent magnitude, a 1.1–3$M_\odot$ mass, a 23.5$D_\odot$ diameter, a 4,554°K effective temperature, a -.54±.12 FeH ratio, and a 619-day rotation. This is a giant star that has already evolved off the main sequence.

The planet, the 105th discovered (2003, La Silla, European Southern Observatory[51]), has a 1.61–2.25AU semi-major axis, a 712 day period, a .20 eccentricity, and a 4.96–9.67M_J minimum mass. The uncertainty here is caused by the uncertainty in the star's mass.

The giant planet dominates the habitable zone, preventing other planets from having stable orbits in the zone. This may be a moot point as the star has only a few tens of millions of years before exhausting its fuel. Any terrestrial-sized moons of the giant planet would have been much colder when the star was on the main sequence, but not necessarily uninhabitable.

The problem lies with the mass of the star. If it is 1.1$M_\odot$ then its main sequence life is only about 4 Gyrs. As its mass goes up, the main sequence period drops rapidly, reaching about 500 million years at a mass of 2$M_\odot$.

Underwood, et al., report that a terrestrial planet orbit would not remain in the habitable zone for at least one billion years.[86]

OGLE-TR-3

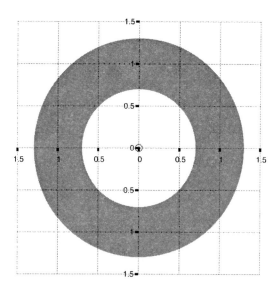

OGLE-TR-3 is a type F9/G0 star in the constellation Scorpius, The Scorpion, its distance is about 9,780 light-years away in the direction of the galactic bulge (RA 17 51 48.95, D -30 13 25.1). It has a 1.0$M_\odot$ mass (with a variance of up to +.3 and down to -.2), a 1.0$R_\odot$ radius (with a variance of up to +.9 and down to -.1), a 6,100±200°K effective temperature, and a metallicity between that of our Sun and -.5 FeH. No apparent magnitude is given.

In addition to being the 106th planet found (2003, European Southern Observatory, [52]), it is one of the hottest and closest to its primary, with a .025AU semi-major axis (2.5 million miles), a 28.5 hour period, an eccentricity of zero, and a .5M_J mass. It is incredibly hot at 2,100°K and has a .68R_J (.17±.01$R_\odot$) radius. Being so close to its primary, the planet is probably face-locked with one side always facing its primary and a large tidally induced bulge. Also, while being only 5$R_\odot$ distant from the primary, it is still a factor of two outside the Roche limit.

It is postulated that the planet was much larger in the past and had spiraled in even closer to the star. But as it got closer it began to lose mass, halting its spiral as it settled at its present position. In spite of its close orbit and high temperature, the planet is stable, although it is probably losing its atmospheric gases as they are accelerated by its proximity to the primary, much like HD 209458.

The habitable zone is unaffected by the presence of the giant planet and the system is old enough to have not only inhabitable planets, but inhabited ones at that.

Currently, there are no orbital simulations available for this star.

HD 73256

HD 73256, is a G8/K0 dwarf in the southern constellation Pyxis, at a distance of about 115 light-years (RA 08 36 23.0155, D -30 02 15.456). It has an apparent magnitude of 8.08, a .98–1.05$M_\odot$ mass, a .89$R_\odot$ radius, a .69$L_\odot$ luminosity, a rotation of 13.9 days, a 5,518±49°K effective temperature, and a .26±.06 FeH ratio. It has an estimated age of 830 million years, and has a correspondingly active chromosphere. It is slightly over luminous for a G8 star, but it is believed that the enhanced metallicity is responsible for that difference.

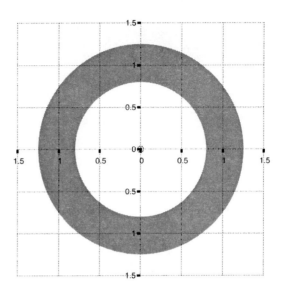

The giant planet, the 107th found (2003, La Silla, European Southern Observatory[53]), has a .037AU semi-major axis, a 2.54863 day period, a .038±.004 eccentricity, and a 1.85M_J minimum mass. It has an estimated dayside temperature of 1,500°K. This is the closest "hot Jupiter" found so far with the radial velocity method.

The habitable zone is unaffected by the presence of the giant planet. As young as the system is, 830 million years, any terrestrial planets are probably still reducing their atmospheres and developing a solid surface.

Underwood, et al., report that a terrestrial planet orbit could remain in the habitable zone for the last one billion years.[86]

HD 104985

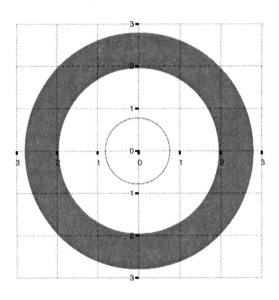

HD 104985 is a G9III giant in the constellation Camelopardalis, The Giraffe, about 332.52 light-years away (RA 12 05 15.1178, D +76 54 20.641). It has an apparent magnitude of 5.797, a 1.5M$_\odot$ mass, a 4,786°K effective temperature, and a -.35 FeH ratio. Notice that although it has a low metallicity compared to our Sun, it still has a giant planet in orbit. How much effect this low metallicity has on terrestrial planet/moon formation is unknown. It could be that only gas giant planets formed in this system.

The giant planet, the 108th found (2003, National Astronomical Observatory, Japan[54]), has a .78AU semi-major axis, a 198.2±.3 day period, a .03±.02 eccentricity, and a 6.3M$_J$ min. mass.

This is a star which has already evolved off the main sequence, after only 2.19Gyrs. What is interesting about this star is that it is actually less luminous now than it was in the beginning! So a planet at the middle of the habitable zone, about 2.4AU, would still be habitable. Unfortunately, for this size star, that still works out to only about 1.5Gyrs. The planet might have simple life, but nothing too complex, perhaps something like trilobites.

The habitable zone is unaffected by the presence of the giant planet.

Underwood, et al., report that a terrestrial planet orbit could remain in the habitable zone for at least one billion years.[86]

HD 10647

HD 10647 is a F8V star in the constellation Eridanus, The River, at a distance of about 56.724 light-years (RA 01 42 29.3157, D -53 44 27.003). It has an apparent magnitude of 5.52, a $1.07\pm.07M_\odot$ mass, a luminosity of $2.51L_\odot$, a $6,143\pm31°K$ effective temperature, a rotation of 7.2 days, an age of 1.75Gyrs, and a -.03±.04 FeH ratio. The star is chromospherically active.

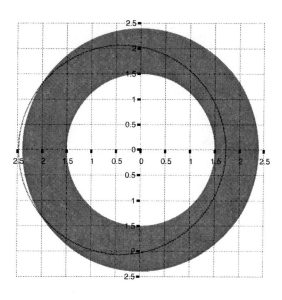

The giant planet, the 109th found (2003, La Silla, European Southern Observatory, Chile[55]), has a 2.1AU semi-major axis, a 1,040±37 day period, a .18±.008 eccentricity, and a $.91M_J$ minimum mass.

Underwood, et al., report that a terrestrial planet orbit could not be confined to the habitable zone for the last one billion years.[86]

Apparent Magnitude = How bright an object appears in the moonless night sky. A 1st magnitude star is roughly 100 times brighter than a 6th magnitude star and thus each magnitude corresponds to an increase of 2.51 times in the star's brightness. This means that a 1st magnitude star is 2.51 times brighter than a 2nd magnitude star. The faintest objects that can be seen with the naked eye have a 6.5 magnitude. Some typical magnitudes are, Sun -26; Full Moon -12.5; First Quarter Moon -10.20; Last Quarter Moon -10.05; Venus -4.6; Jupiter, Mars -2.9; Sirius -1.5; Naked eye limit 6.5; Binocular limit 10.

HD 41004A

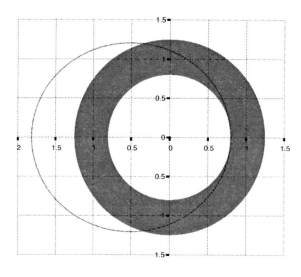

HD 41004A is part of a binary system whose partner also has an orbiting brown dwarf. It is a K1V/K2V dwarf in the constellation Pictor, The Easel, about 140.18 light-years away (RA 05 59 49.6493, D -48 14 22.889). HD 41004A has an apparent magnitude of 8.65, a $.7M_\odot$ mass, a $.65L_\odot$ luminosity, a 5,085°K effective temperature, a 27 day rotational period, an estimated age of 1.6Gyrs, and a -.0±.1 FeH ratio.

The giant planet, the 110th found (2003, La Silla, European Southern Observatory, Chile[55]), has a 1.31AU semi-major axis, a 655±37 day period, a .39±.17 eccentricity, and a $2.3M_J$ mini. mass.

The partner star is an M4V star separated from the primary by a visual distance of .5 arcseconds (known as a visual binary). This is a distance of about 1.2 light-years (depending on the accuracy of the distance from Earth). It is orbited by a $19M_J$ brown dwarf. Orbital parameters are unknown for both objects.

Underwood, et al., report that a terrestrial planet orbit could only be confined to part of the habitable zone for the last one billion years.[86]

HD 65216

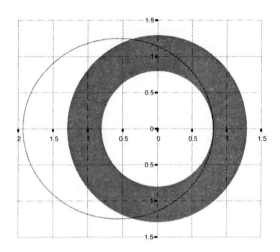

HD 65216 is a G5V dwarf in the constellation Carinae, The Keel (of the ship), at a distance of about 111.818 light-years away (RA 07 53 41.3223, D-63 38 50.363). It has an apparent magnitude of 7.97, a .94±.02M$_\odot$ mass, a .71L$_\odot$ luminosity, a 5,666±31°K effective temperature, and a -.12±.04 FeH ratio.

The giant planet, the 111th found (2003, La Silla, European Southern Observatory, Chile[55]), has a 1.37AU semi-major axis, a 613±11.4 day period, a .41±.06 eccentricity, and a 1.21M$_J$ minimum mass.

Underwood, et al., report that a terrestrial planet orbit would not remain in the habitable zone for at least one billion years.[86]

HD 111232

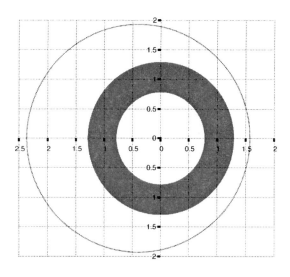

HD 111232 is a G5V dwarf in the constellation Musca, The Fly, at a distance of about 94.54 light-years away (RA 12 48 51.7543, D -68 25 30.544). It has an apparent magnitude of 7.61, a .78$M_\odot$ mass, a .69$L_\odot$ luminosity, a 5,494°K effective temperature, a 30.7 day rotation, an apparent age of 5.2Gyrs, and a -.36±.04 FeH ratio. This is a high velocity star (104.5km/sec) and probably belongs to the thick disk population of stars. The Hipparcos catalog lists it as a possible binary, but no companion (with less than a 3 km/sec relative velocity) is visible by speckle interferometry.

The giant planet, the 112th found (2003, La Silla, European Southern Observatory, Chile[55]), has a 1.97AU semi-major axis, a 1143±14 day period, a .20±.01 eccentricity, and a 6.8M_J minimum mass.

Underwood, et al., report that a terrestrial planet orbit would only remain in the habitable zone for at least one billion years.[86]

HD 142415

HD 142415 is a G1V dwarf in the constellation Triangulum Australe, The Southern Triangle, at a distance of 112.686±1.2 light-years (RA 15 57 40.7907, D -60 12 00.926). It has an apparent magnitude of 7.34, a .1.03$M_\odot$ mass, a 1.14$L_\odot$ luminosity, a 6,045±44°K effective temperature, a 9.6 day rotation, an estimated age of 1.1Gyrs, and a .21±.05 FeH ratio. A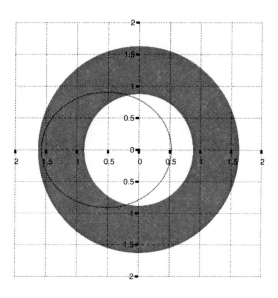
Rosat-All-Sky-Survey X-ray source (1RXSJ155740.7–601154) is observed within five arcseconds of this star, but no bright companions are observed in visible light range.

The giant planet, the 113th found (2003, La Silla, European Southern Observatory, Chile[55]), has a 1.05AU semi-major axis, a 386±1.6 day period, a .50 eccentricity (fixed), and a 1.62M_J minimum mass. Because the orbital period is close to one year, the observations made tend to undersample the planet's closest approach to its star. As a result the eccentricity has been fixed to .5, but any value between .2 and .8 would satisfy the data accumulated over the last four years. (Its closest approach coincided with being hidden by daylight).

Underwood, et al., report that a terrestrial planet orbit would not remain in the habitable zone for at least one billion years.[86]

HD 216770

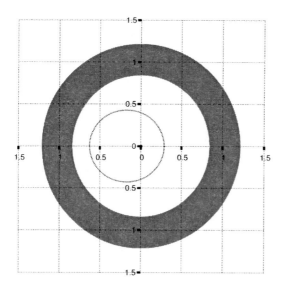

HD 216770 is a K1V type star in the constellation Piscis Austrinus, The Southern Fish, at a distance of 123.5±.03 light-years (RA 22 55 53.7097, D -26 39 31.547). It has an apparent magnitude of 8.1, a .90$M_\odot$ mass, a .79$L_\odot$ luminosity, a 5,423±41°K effective temperature, a 35.6 day rotation, a 3.1Gyr estimated age, and a .26±.04 FeH ratio.

The giant planet, the 115th found (2003, La Silla, European Southern Observatory, Chile[55]), has a .46AU semi-major axis, a 118.45±.4 day period, a .37±.06 eccentricity, and a .65M_J minimum mass.

Underwood, et al., report that a terrestrial planet orbit could remain in the habitable zone for the last one billion years.[86]

HD 70642

HD 70642 (HIP 40952) is a G5IV-V type star in the constellation Puppis, The Stern (of the Ship), at a distance of 123.88 light-years (RA 08 21 28.1361, D -39 42 19.474). It has an 7.18 apparent magnitude, a $1.0\pm.05M_\odot$ mass, and a .16±.042FeH ratio. The star is chromospherically inactive with an inferred age of 4Gyrs. Its 5,670°K effective temperature seems to support its status as a middle-aged G5 dwarf star.

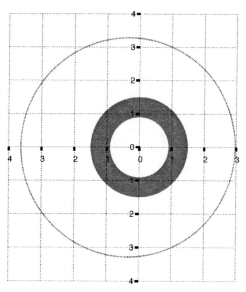

The giant planet, the 116th found (2003, Anglo-Australian Telescope[56]), has a 3.3AU semi-major axis, a 2,231±400 day period, a .10±.06 eccentricity, and a $2.0M_J$ minimum mass. This is one of the planets in a nearly circular orbit.

Terrestrial-sized planets could have safe orbits in the habitable zone, but the closer such planets are to the inner edge of the zone the less influence the giant planet would have on their orbit.

Underwood, et al., report that a terrestrial planet orbit could remain in the habitable zone for the last one billion years.[86] In addition, like our own solar system, the giant planet remains outside the habitable zone for the star's entire main sequence life.

HD 330075

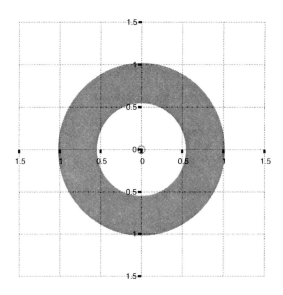

HD 330075 is a G5 type dwarf star in the constellation Norma, The Ruler, at a distance of about 163.655 light-years (RA 15 49 37.6913, D -49 57 48.692). It has an 9.36 apparent magnitude, a $.95M_\odot$ mass, a $.47L_\odot$ luminosity, 4,973°K effective temperature, and an unknown FeH ratio.

The giant planet, the 117th found (2003, La Silla, European Southern Observatory[57]), has a .044AU semi-major axis, a 3.369±.004 day period, a 0.0 eccentricity (fixed, that is assumed to be), and a $.76M_J$ minimum mass.

With the giant planet so close to the primary it would have little effect on terrestrial-sized planets in the habitable zone.

Currently, there are no orbital simulations available for this star.

HD 59686

HD 59686 is a K2III type giant star in the constellation Gemini, The Twins, at a distance of about 299.92 light-years (RA 07 31 48.3969, D +17 05 09.765). It has an 5.45 apparent magnitude, and a 2.6M$_\odot$ mass, and an unknown FeH ratio. It is receding from our Sun at a rate of 40 km/sec. From its apparent magnitude and distance, the luminosity should be 47L$_\odot$.

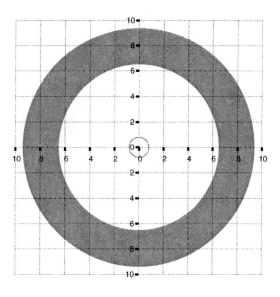

The giant planet, the 118th found (2003, Lick Observatory[58]), has a .8AU semi-major axis, a 303 day period, an unknown eccentricity, and a 6.5M$_J$ min. mass.

The giant planet is sufficiently far from the habitable zone to have no effect on any planets orbiting there. The fact that this is a giant star that has evolved off the main sequence means the habitable zone has likewise moved outward from its initial main-sequence position. Habitable planets have thus only recently thawed out.

Underwood, et al., report that a terrestrial planet orbit would have remained in the previous habitable zone for at least one billion years.[86]

HD 219449 (91 Aqr)

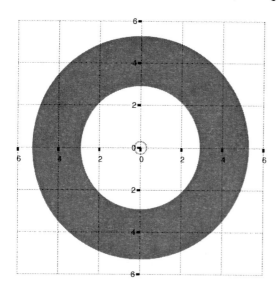

HD 219449 (91 Aqr) is a type K0III giant star in the constellation Aquarius, The Water Carrier, at a distance of about 146.7 light-years (RA 23 15 53.4947, D -09 05 15.853). It has an apparent magnitude of 4.21. Based on its apparent magnitude and distance, it has between 1.895 and 2.447$M_\odot$ mass, and between 12.897 and 35.81$L_\odot$ luminosity. Such a wide latitude in luminosity makes reasonable estimates of the habitable zone impossible.

The giant planet, the 119th found (2003, Lick Observatory[58]), has a .3AU semi-major axis, a 182 day period, an unknown eccentricity, and a 2.9M_J min. mass.

At its lower mass estimate, the star spent only 900 million years on the main sequence and will spend only a few million years in the giant star stage. At the higher estimate it spent under 344 million years, and will spend less than a million years as a giant star.

For the purposes of this chart above, a luminosity of 13$L_\odot$ is assumed, as is an eccentricity of zero.

Underwood, et al., report that a terrestrial planet orbit would not remain in the habitable zone for the last one billion years.[86] However, when the star was in its main sequence phase of life, a terrestrial planet orbit would have remained within the previous habitable zone for at least one billion years.

OGLE-TR-113

OGLE-TR-113 is a type G star in the constellation Carinae, The Keel (of the Ship), its distance is about 4,890 light-years away (RA 10 52 24.40, D -61 26 48.5). It has a 14.42 apparent magnitude, a $.79\pm.06M_\odot$ mass, a $.78\pm.06R_\odot$ radius, a 4,800°K effective temperature, and a $.17\pm.14$ FeH metallicity.

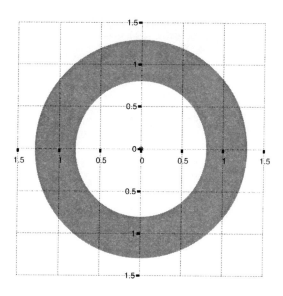

This is the 120th found (Paranal, European Southern Observatory, Chile[59]), it is one of the hottest and closest to its primary, with a $.02299\pm.00058$AU semi-major axis, a 34.794 ($\pm.39744$ seconds) hour period, a zero eccentricity, a $1.35\pm.22M_J$ mass, a $1.08^{+.07}_{-.05}R_J$ ($.17\pm.01R_\odot$) radius, a 1,144°K effective temperature. The uncertainty in the planet's mass and radius are due entirely to the uncertainty in the primary's mass. Being so close to its primary, the planet is probably face-locked with one side always facing its primary and a large tidally induced bulge. Also, while being only $5R_\odot$ distant from the primary, it is still a factor of two outside the Roche limit. The orbital inclination is $1.6°\pm2.2$ degrees from our line of sight.

It is postulated that the planet was much larger in the past and had spiraled in even closer to the star. But as it got closer it began to lose mass, halting its spiral as it settled at its present position. In spite of its close orbit and high temperature, the planet is stable, although it is probably losing its atmospheric gases as they are accelerated by its proximity to the primary, much like HD 209458.

The habitable zone is unaffected by the presence of the giant planet and the system is old enough to have not only inhabitable planets, but inhabited ones at that.

OGLE-TR-132

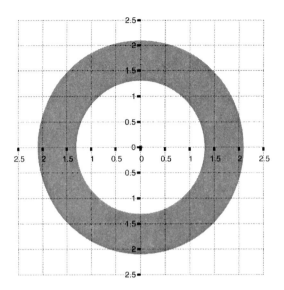

OGLE-TR-132 is a type G star in the constellation Carinae, The Keel (of the Ship), its distance is about 4,890 light-years away in the direction of the galactic bulge (RA 10 50 34.72, D -61 57 25.9). It has a 15.72 apparent magnitude, a 1.34±.10M$_\odot$ mass, and a .43±.10 FeH metallicity.

In addition to being one of the furthest planets discovered, and the 121st (Paranal, European Southern Observatory, Chile[59]), it is one of the hottest and closest to its primary, with a .0306±.0008AU semi-major axis (2.8 million miles), a 40.556 hour period, an eccentricity of zero, a 1.19±.13M$_J$ mass, a 1.13±.08R$_J$ radius, and a 1,821 effective temperature. Being so close to its primary, the planet is probably face-locked with one side always facing its primary and a large tidally induced bulge. Also, while being only 5R$_\odot$ distant from the primary, it is still a factor of two outside the Roche limit.

It is postulated that the planet was much larger in the past and had spiraled in even closer to the star. But as it got closer it began to lose mass, halting its spiral as it settled at its present position. In spite of its close orbit and high temperature, the planet is stable, although it is probably losing its atmospheric gases as they are accelerated by its proximity to the primary, much like HD 209458.

The habitable zone is unaffected by the presence of the giant planet and the system is old enough to have not only inhabitable planets, but inhabited ones at that.

Currently, there are no orbital simulations available for this star.

HD 37605

HD 37605 is a type K0V star in the constellation Orion, The Hunter. Its distance is about 139.794 light-years away (RA 05 40 01.7296, D +06 03 38.085). It has an 8.69 apparent magnitude, 5.51 absolute magnitude, a .851$M_\odot$ mass, 5,475±50°K effective temperature, and a .39±.06 FeH metallicity. This appears to be an older star with little chromospheric activity.

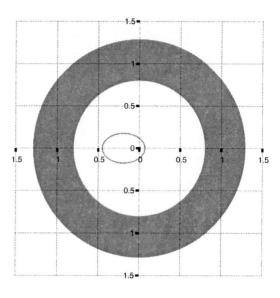

The giant planet, the 122nd found (2004, Hobby-Eberly Telescope[60]), has a .26±.01AU semi-major axis, a 54.2±.23 day period, a .736±.0095 eccentricity, and a 2.85±.26M_J minimum mass.

The habitable zone is unaffected by the presence of the giant planet and the system is old enough to have not only inhabitable planets, but inhabited ones at that.

Currently, there are no orbital simulations available for this star.

2M1207

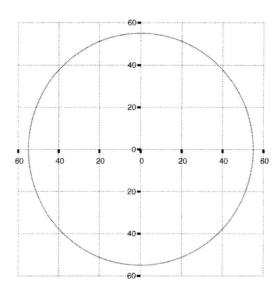

2M1207 (2MASS-WJ1207334–393254) is a type M8 brown dwarf star in the constellation Centaurus, The Centaur, at a distance of about 228 light-years (RA12 07 33.4, D -39 32 54). This failed star glows only by the energy left over from its initial burning of deuterium in its core when it first formed. Based on its motion and spectrum it appears to be a member of the TW Hydrae Association, which formed about 8–10 million years ago. This star's age is estimated at 8^{+4}_{-3} Myrs. It has a low surface gravity, indicative of a very young object, 25M_J mass. The fact that this star has already ceased fusing hydrogen means its habitable zone is nonexistent. Any planets will be warm only from their own internal heating processes of radioactive decay and compression. Life of any kind is highly improbable.

The giant planet, the 123rd found (2004, Paranal, European Southern Observatory, Chile[61]), is at a distance of 55AU, with an unknown period, an unknown eccentricity, and a 5±2M_J minimum mass. Because its primary is so dim, an actual measurement of the energy given off by the giant planet can be estimated, with a spectral type of L5 – L9.5, and an effective temperature of 1,250±200°K.

If this giant planet is indeed orbiting 2M1207, then it will be the first planet found by direct imaging instead of radial velocity or primary transit. There is a small chance (9x10–8) that this is a "field" object, something between a small L8 brown dwarf star at 187 light-years or an L5 dwarf at 420 light-years.

TrES-1

TrES-1 is a type K0V star in the constellation Lyra, The Lyre, at a distance of about 511.82 light-years (RA 19 04 09.8, D +36 37 57). It has an apparent magnitude of 11.79, 5,214±23°F effective temperature, and a .001±.03 FeH metallicity. Based on its apparent magnitude and distance, it has a .87±.03$M_\odot$ mass, a .83±.03$R_\odot$ radius, and .18$L_\odot$ luminosity.

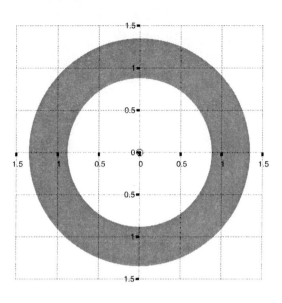

The giant planet, the 124th found (2004, Tenerife in Canary Islands, Lowell Observatory, Mt. Palomar[62]), has a .0393±.0011AU semi-major axis, a 3.030065±8×10^6 day period, an unknown eccentricity, a .729±.036M_J minimum mass, and a 1.08±.05R_J radius. Because this planet was detected via transit, the ratio of the radius of the planet to that of its primary is $0.130^{+.009}_{-.003}$. With an orbit of just four million miles from its primary, the planet has a surface temperature of 1,500°K.

Currently, there are no orbital simulations available for this star.

OGLE-TR-111

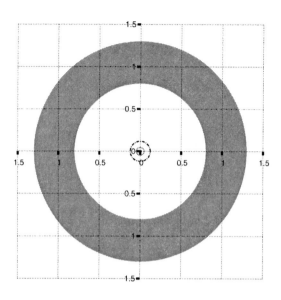

OGLE-TR-111 is a late-G or early K type star in the constellation Carinae, The Keel (of the Ship), its distance is about 2,657 light-years away in the direction of the galactic bulge (RA 10 53 17.91, D -61 24 20.3). It has a 15.55 apparent magnitude (6.82 absolute magnitude), a $.82^{+.15}_{-.02} M_\odot$ mass, $.85^{+.1}_{-.03}$ radius, 5,070±400°K temperature, and a .12±.28 FeH metallicity.

This is the first planet discovered that straddles the boundary between planets found via OGLE (periods less than 2.5 days) and those found via radial velocity (periods over 2.5 days). It was the 126th planet found (2004, Paranal, European Southern Observatory[64]), with a .047±.001AU semi-major axis, a 4.01610 day period (fixed), a zero eccentricity, a .53±.11M_J mass, a $1.00^{+.13}_{-.06} R_J$ radius, 904°K temperature, and an apparent density of $.61^{+.39}_{-.26}$ g/cm^3. Being so close to its primary, the planet is probably face-locked with one side always facing its primary and a large tidally induced bulge.

Further analysis of the data has tentatively identified a second planet in this system, the 144th (2005, La Silla, European Southern Observatory, Chile[73]), with a .12±.01AU semi-major axis, a 16.0644±.005 day period, eccentricity of zero, and a .7±.2M_J mass. It has a .85±.15R_J radius, and an apparent density of 1.4±.3 g/cm^3. This is a tentative identification because the period of the second planet is an integer multiple of the inner planet's orbit. This is not unusual, as Jupiter's moons Io and Ganymede share a 1:4 resonance.

The habitable zone is unaffected by the presence of either giant planet and the system is old enough to have not only inhabitable planets, but inhabited ones at that.

Currently, there are no orbital simulations available for this star.

Ross 905, GJ 436

Ross 905 (Gliese 476, HIP 57087) is one of the stars in this book without a Henry Draper number as an index. It is a small, dim, red, type M2.5V star, located about 33.357 light-years away in the constellation of Leo, (RA 11 42 11.0941, D +26 42 23.652), The Lion, with a 10.67 apparent magnitude, a .41±.05$M_\odot$ estimated mass, a .38$D_\odot$ diameter, a .025$L_\odot$ luminosity, and a FeH ratio of zero, the same as our Sun. The star appears to be about 3 Gyrs old.

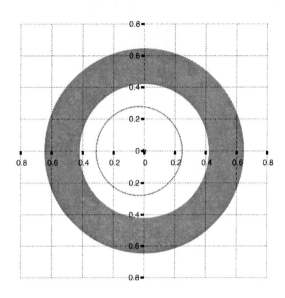

This was the second discovery of a planet around an M dwarf by radial velocity measurements, and only the second out of 150 M-type stars studied so far.

The planet, the 128th discovered (2004, Keck Observatory[66]), has a .0278AU semi-major axis, a 2.6441±.0005 day period, a .12 eccentricity, and a .067±.007M_J minimum mass (1.2$M_{Neptune}$, 21M_{Earth}). Its estimated temperature is 620°K. There are questions as to the composition of this giant planet, e.g., primarily Hydrogen/Helium gases, ice/rock, or rock dominated.

Orbital simulations are not yet avaiable for this system.

HD 88133

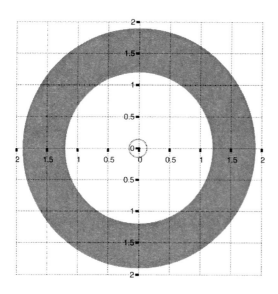

HD 88133 is a type G5 IV giant star in the constellation Leo, The Lion, at a distance of about 242.74 light-years (RA 10 10 07.6754, D +18 11 12.736). It has an 8.0 apparent magnitude. Based on its apparent magnitude and distance, it has a 1.2$M_\odot$ mass, 1.93$R_\odot$ radius, a 5,494±23°K effective temperature, a rotation of 48 days, and 2.8$L_\odot$ luminosity. It has a .34±.04 Fe/H metallicity ratio. From its radius and luminosity, this is an evolved giant star.

The giant planet, the 129th found (2004, Keck Observatory[67]), has a .046AU semi-major axis, a 3.415±.001 day period, a .11±.05 eccentricity, and a .29M_J minimum mass. If the planet has a core, it should have a .97R_J radius. If not, then the radius would be 1.12R_J.

Because this is an evolved star, the habitable zone has left behind any inner planets that might have harbored life while it was on the main sequence. Planets formerly too cool for habitability are now warm enough for liquid water on their surfaces.

Currently, there are no orbital simulations available for this star.

HD 102117

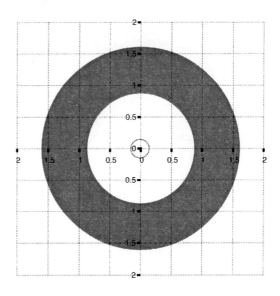

HD 102117 is a type G6V star in the constellation Centaurus, The Centaur, at a distance of about 194.048 light-years (RA 11 44 50.4616, D -58 42 13.354). The distance given is from the paper of the discoverers, with a 16.8±.7mas parallax measurement acquired from Hippaarcos. SIMBAD lists a 23.81mas measurement, for a distance of 136.92 light-years. As you can see this is a difference of almost 60 light-years! This could have a dramatic impact on the orbital elements of this system, as well as the habitability zone. The star has reached the end of its hydrogen-burning cycle and is burning helium, hence its larger than normal radius.

It has a .95±.05$M_\odot$ mass, 5,429±20°K effective temperature, a .04±.08 Fe/H metallicity, a 1.75±27$R_\odot$ radius, and 1.03±.08$L_\odot$ luminosity. Its estimated age is 10Gyrs. It is chromospherically very quiet and photometrically stable.

The giant planet, the 130th found (2004, Anglo-Australian Telescope[68]), has a .15±.01AU semi-major axis, a 20.8±.1 day period, a .08±.05 eccentricity, and a .18±.03M_J minimum mass.

The habitable zone has moved beyond the orbits of any habitable planets from this star's main sequence life; hence any planets in the habitable zone shown above have only recently become habitable.

Currently, there are no orbital simulations available for this star.

HD 117618

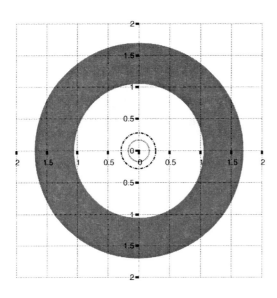

HD 117618 is a type G2V star in the constellation Centaurus, The Centaur, at a distance of about 123.95 light-years (RA 11 44 50.4616, D -58 42 13.354). It has a $1.05\pm.05M_\odot$ mass, $5,855\pm20°K$ effective temperature, a $.04\pm.08$ Fe/H metallicity, and $1.34\pm.35L_e$ luminosity. Its estimated age is 4.8–6.3 Gyrs. It is chromospherically quiet and photometrically stable.

The preliminary data from this star yield two possible orbital fits for its giant planet, the 131st found (2004, Anglo-Australian Telescope[68]). The 25.8-day orbit seems to satisfy the data the best, but that leaves an unexplained peak at 52.2 days. Quite possibly there are two giant planets orbiting this star and further study is required to refine the data.

The inner orbital fit has a $.17\pm.01$AU semi-major axis, a $25.8\pm.5$ day period, a $.37\pm.01$ eccentricity, and a $.16\pm.03M_J$ minimum mass.

The outer orbital fit has a $.28\pm.02$AU semi-major axis, a $52.2\pm.5$ day period, a $.39\pm.01$ eccentricity, and a $.19\pm.04M_J$ minimum mass.

Both orbits are well inside the habitable zone, leaving it available for possible terrestrial planets. Both are plotted above.

Currently, there are no orbital simulations available for this star.

HD 208487

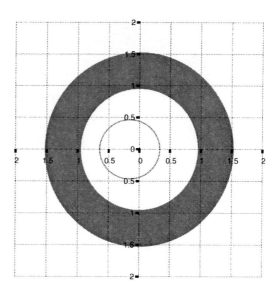

HD 208487 is a type G2V star in the constellation Grus, The Crane, at a distance of about 194.048 light-years (RA 21 57 19.8477, D -37 45 49.037). It has a $.95\pm.05M_\odot$ mass, $5,929\pm20°K$ effective temperature, a $.06\pm.05$ Fe/H metallicity, and $1.07\pm.065L_\odot$ luminosity. Its estimated age is 6.3–10 Gyrs. It is chromospherically quiet and photometrically stable.

The giant planet, the 132nd found (2004, Anglo-Australian Telescope[68]), has a $.49\pm.04$AU semi-major axis, a 130 ± 1 day period, a $.32\pm.1$ eccentricity, and a $.45\pm.05\ M_J$ minimum mass.

Currently, there are no orbital simulations available for this star.

HD 154857

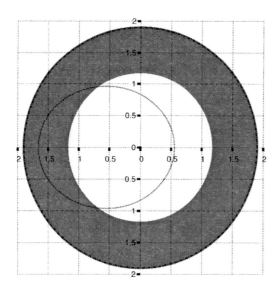

HD 154857 is a type G5 star in the constellation Grus, The Crane, at a distance of about 223.44 light-years (RA 17 11 15.7219, D -56 40 50.865). Its absolute magnitude is about 2 magnitudes higher than expected, suggesting the star is evolving. It has a $1.17\pm.05M_\odot$ mass, $5,628\pm40°K$ effective temperature, a $-.23\pm.03$ Fe/H metallicity, and $3.91\pm1.14L_\odot$ luminosity. Its estimated age is 5 Gyrs. It is chromospherically quiet, like most evolved stars moving towards subgiant status.

The giant planet, the 133rd found (2004, Anglo-Australian Telescope[69]), has a 1.11 AU semi-major axis, a 398.5 day period, a .51 eccentricity, and a 1.80 M_J min. mass.

There is evidence in the data that a second planet, the 134th planet found (2004, Anglo-Australian Telescope[69]), orbits this star, with a mass greater than 1 M_J. and a period over two years.

For the purposes of the chart above, an eccentricity of zero has been assumed, as well as a minimum period of three years and a 1.9 AU semi-major axis.

Currently, there are no orbital simulations available for this star.

OGLE-TR-10

OGLE-TR-10 is a type G2V star in the constellation Sagittarius, The Archer, and is located 4,890 light-years away (RA 17 51 28.25, D -29 52 34.9). In fact, it is not even in our arm of the galaxy, the Orion Arm, but is in the Sagittarius Arm. It is very similar to our Sun, with a $1.00\pm.05M_\odot$ mass, a $1.0\pm.1R_\odot$ radius, a $5,750\pm100°K$ effective temperature, a 14.9 apparent magnitude, a $0.0\pm.2$ Fe/H metallicity, and a 4 Gyr age estimate.

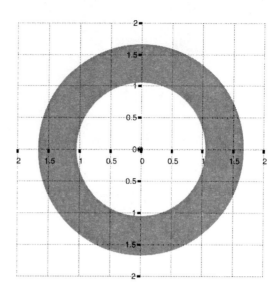

It is the 136th planet found (2004, Paranal, European Southern Observatory, Chile[70]), with a $.04162\pm.00069$ AU semi-major axis, a $3.101386\pm.00003$ day period, an eccentricity of zero, and a $.57\pm.3M_J$ mass. It is incredibly hot at 1,900°K, has a $1.24\pm.09R_J$ radius, and a density of $.38\pm.1$ g/cm^3. Being so close to its primary, the planet is probably face-locked with one side always facing its primary and a sizeable tidal bulge.

It is postulated that the planet was much larger in the past and had spiraled in even closer to the star. But as it got closer it began to lose mass, halting its spiral and reversing it as it lost mass until it reached its present position. In spite of its close orbit and high temperature, the planet is stable, although it is losing its atmospheric gases as they are excited by its proximity to the primary, much like HD 209458. In spite of the mass loss, the planet will most likely easily survive until its primary reaches nova stage.

The habitable zone is unaffected by the presence of the giant planet and the system is old enough to have not only inhabitable planets, but inhabited ones at that.

Currently, there are no orbital simulations available for this star.

HD 202206

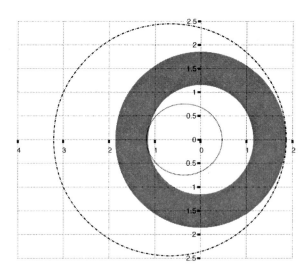

HD 202206 is a type G6V star in the constellation Capricorn, The Goat, at a distance of about 151 light-years (RA 21 14 57.7693, D -20 47 21.154). It has an absolute magnitude .4 higher than expected, and a higher effective temperature, probably caused by the high metallicity. It has a 1.15 $M_\odot$ mass, a 5,765±40°K effective temperature, a .37±.07 Fe/H metallicity, and 1.07$L_\odot$ luminosity. Its estimated age is 5.6±1.2 Gyrs.

This system is unusual as it calls into question current theories of planetary system evolution. The inner object is a brown dwarf (failed star). If it is assumed that the inner object formed concurrently with its primary, then it will be hard to explain the presence of the outer planet. Either the outer planet was also formed concurrently, in an accretion disk around the binary composed by the main star and the brown dwarf, or it formed in the stellar protoplanetary disk left after the other two were formed. If the latter, then the implication is that the inner planet also formed in the protoplanetary disk, and therefore is not a brown dwarf. This last assumption leads to the conclusion that protoplanetary disks may be much more massive than thought.

The giant planet, the 137th found (2004, La Silla, European Southern Observatory[71]), has a 2.5420 AU semi-major axis, a 1383.4±18.4 day period, a .26692 eccentricity, and a 2.43653 M_J minimum mass.

Careful analysis of the data, solving for a stable system, gives the conclusion that the orbits of the brown dwarf and the giant planet are not coplanar. The giant planet orbits at an inclination of 105.68256° to the brown dwarf's orbit.

HD 45350

HD 45350 is a type G5V star in the constellation Auriga, The Charioteer, at a distance of about 159.56 light-years (RA 06 28 45.7103, D +38 57 46.667). It has an absolute magnitude .8 higher than expected, and a higher effective temperature, probably caused by the high metallicity. It has a 1.02±.1 $M_\odot$ mass, a 5,616°K effective temperature, a .29±.07 Fe/H metallicity, and 1.02$L_\odot$ luminosity. It is chromospherically quiet with a rotation of 39 days. Its estimated age is 6–10 Gyrs.

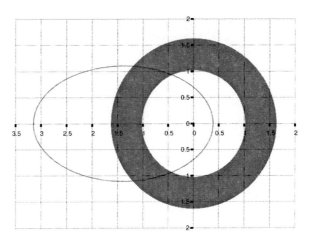

The giant planet, the 138th found (2004, Keck Observatory[72]), has a 1.77 AU semi-major axis, a 890.76±37.42 day period, a .78±.09 eccentricity, and a .98 M_j minimum mass.

The giant planet crosses through the habitable zone and probably precludes any chances that there is a habitable planet.

HD 99492

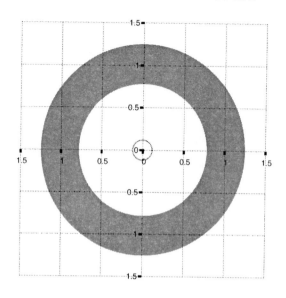

HD 99492 (83 Leo B) is a type K2V star in the constellation Leo, The Lion, at a distance of about 159.56 light-years (RA 11 26 46.2771, D +03 00 22.781). It has an absolute magnitude .3 higher than expected, and a higher effective temperature, probably caused by the high metallicity. It has a .775±.08 $M_\odot$ mass, a 4,954°K effective temperature, a .36 Fe/H metallicity, and .259 $L_\odot$ luminosity. It is chromospherically average with a rotation of 45 days. Its estimated age is 2–6 Gyrs.

The giant planet, the 139th found (2004, Keck Observatory[72]), has a .12 AU semi-major axis, a 17.038±.011 day period, a .05±.12 eccentricity, and a .11 M_J minimum mass.

Currently, there are no orbital simulations available for this star. However, the giant planet is far enough from the habitable zone to have no effect on any planets orbiting there.

HD 117207

HD 117207 is a type G8IV/V star in the constellation Centaurus, The Centaur, at a distance of about 107.62 light-years (RA 13 29 21.113, D -35 34 15.589). It has an absolute magnitude .55 higher than expected, and a higher effective temperature, probably caused by the high metallicity. It has a 1.04±.1 $M_\odot$ mass, a 5,723°K effective tem-

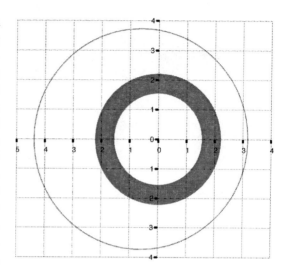

perature, a .27 Fe/H metallicity, and 1.12$L_\odot$ luminosity. It is chromospherically quiet with a rotation of 36 days. Its estimated age is 6–10 Gyrs.

The giant planet, the 140th found (2004, Keck Observatory[72]), has a 3.78 AU semi-major axis, a 2,627.08±63.51 day period (7.19±.3 years), a .16±.08 eccentricity, and a 2.06 M_j minimum mass.

Currently, there are no orbital simulations available for this star.

HD 183263

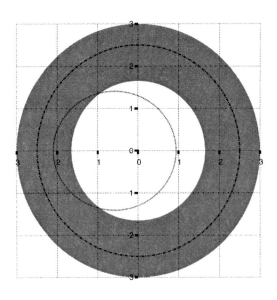

HD 183263 is a type G2IV star in the constellation Aquila, The Eagle, at a distance of about 172.2 light-years (RA 19 28 24.5727, D +08 21 28.995). It has an absolute magnitude .4 higher than expected, and a higher effective temperature, probably caused by the high metallicity. It has a 1.17±.1 $M_\odot$ mass, a 5,936°K effective temperature, a .3 Fe/H metallicity, and 1.12$L_\odot$ luminosity. It is chromospherically quiet with a rotation of 32 days. Its estimated age is 6–10 Gyrs and nearing the end of its main sequence life.

The giant planet, the 141st found (2004, Keck Observatory[72]), has a 1.52 AU semi-major axis, a 634±5 day period (7.19±.3 years), a .38 eccentricity, and a 3.69 M_J minimum mass.

The data indicates that there is a more distant companion 142nd, (2004, Keck Observatory[72]) with a period of more than 4 yrs, semi-major axis of at least 2.5 AU, and a probable greater-than 4 M_J mass. Hipparcos data doesn't detect a wobble in the primary's movement, so the companion must have a 50 M_J mass, or less, or a period greater than the 8-year observational time of the Hipparcos mission.

Currently, there are no orbital simulations available for this star. However, with the inner planet crossing into the habitable zone it probably precludes the possibility of a habitable planet in the system.

One the other hand, if there is indeed a second giant planet, and its orbit is not more than 3AU from the primary, it could have a terrestrial-sized habitable moon.

HD 188015

HD 188015 is a type G5IV star in the constellation Vulpecula, The Fox, at a distance of about 171.5 light-years (RA 19 52 04.5435, D +28 06 01.356). It has an absolute magnitude .5 higher than expected, and a higher effective temperature, probably caused by the high metallicity. It has a 1.08±.1 $M_\odot$ mass, a 5,745°K effective temperature, a .29 Fe/H metallicity, and 1.41$L_\odot$ luminosity. It is chromospherically quiet with a rotation of 36 days. Its estimated age is 6–10 Gyrs.

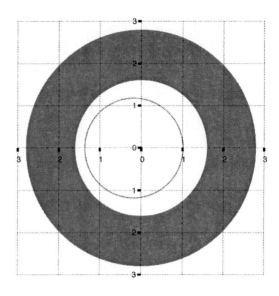

The giant planet, the 143rd found (2004, Keck Observatory[72]), has a 1.19 AU semi-major axis, a 456.5 day period, a .15 eccentricity, and a 1.26 M_j min. mass.

Currently, there are no orbital simulations available for this star. A habitable planet could probably survive in the system as long as it was close to the outer perimeter of the habitable zone.

HD 93083

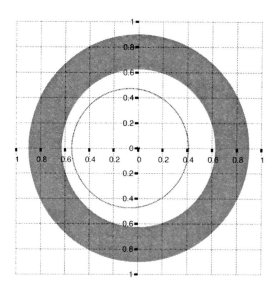

HD 93083 is a type K3V (although SIMBAD lists it as a K2V) star in the constellation Antila, The Ant, at a distance of about 94.2 light-years (RA 10 44 20.9149, D -33 34 37.279). It has a .70±.04 $M_\odot$ mass, a 4,995±50°K effective temperature, a .15±.06 Fe/H metallicity, and .41$L_\odot$ luminosity. It is chromospherically quiet with a rotation of 48 days.

The giant planet, the 145th found (2005, La Silla, European Southern Observatory, Chile[74]), has a .477 AU semi-major axis, a 143.58±.60 day period, a .14±.03 eccentricity, and a .37 M_J Minimum mass, slightly larger than Saturn's .298M_J mass.

Currently, there are no orbital simulations available for this star, but with the giant planet so close to the habitable zone and the habitable zone so narrow, a terrestrial planet might not survive long enough to develop a biosphere.

HD 101930

HD 101930 is a type K1V star in the constellation Centaurus, The Centaur, at a distance of about 99.4 light-years (RA 11 43 30.1115, D -58 00 24.793). It has a .74±.05 $M_\odot$ mass, a 5,079±62°K effective temperature, a .17±.06 Fe/H metallicity, and .49$L_\odot$ luminosity. It is chromospherically quiet with a rotation of 46 days.

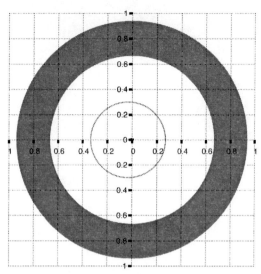

The giant planet, the 146th found (2005, La Silla, European Southern Observatory, Chile[74]), has a .302 AU semi-major axis, a 70.46±.18 day period, a .11±.02 eccentricity, and a .30 M_J minimum mass, slightly larger than Saturn's .298M_J mass.

Currently, there are no orbital simulations available for this star, but with the giant planet so close to the primary a terrestrial planet could probably survive in the outer portion of the habitable zone.

HD 2638

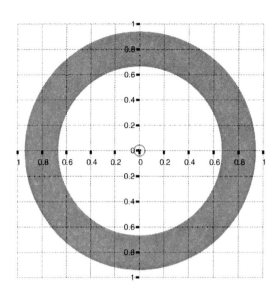

HD 2638 is a type G5 star in the constellation Cetus, The Whale, at a distance of about 175.09 light-years (RA 00 29 59.8720, D -05 45 50.406). It has a .93±.1 $M_\odot$ mass, a 5,192±38°K effective temperature, a .16±.05 Fe/H metallicity, and .47$L_\odot$ luminosity. It is chromospherically quiet with a rotation of 37 days.

The giant planet, the 147th found (2005, La Silla, European Southern Observatory, Chile[75]), has a .044 AU semi-major axis, a 3.4442±.0002 day period, a zero eccentricity, and a .48 M_J min. mass, close to Saturn's .298M_J mass.

Currently, there are no orbital simulations available for this star, but with the giant planet so close to the primary a terrestrial planet could probably survive anywhere in the habitable zone.

HD 27894

HD 27894 is a type K2V star in the constellation Reticulum, The Net, at a distance of about 138.1 light-years (RA 04 20 47.0473, D -59 24 39.014). It has a .75 $M_\odot$ mass, a 4,875±81°K effective temperature, a .03±.07 Fe/H metallicity, and .36$L_\odot$ luminosity.

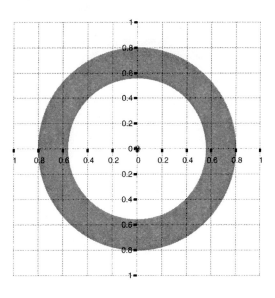

The giant planet, the 148th found (2005, La Silla, European Southern Observatory, Chile[75]), has a .122 AU semi-major axis, a 17.991±.007 day period, a .049±9.7 eccentricity, and a .62 M_J min. mass, slightly more than double Saturn's .298M_J mass.

Currently, there are no orbital simulations available for this star, but with the giant planet so close to the primary a terrestrial planet could probably survive anywhere in the habitable zone.

HD 63454

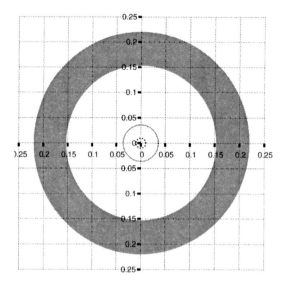

HD 63454 is a type K4V star in the constellation Chamaeleon, The Chameleon, at a distance of about 116.7 light-years (RA 07 39 21.8511, D -78 16 44.300). It has a .8 $M_\odot$ mass, a 4,841±65°K effective temperature, a .11±.07 Fe/H metallicity, and .26$L_\odot$ luminosity.

The giant planet, the 149th found (2005, La Silla, European Southern Observatory, Chile[75]), has a .036 AU semi-major axis, a 2.81782±.000095 day period, a zero eccentricity, and a .38 M_J min. mass, slightly larger than Saturn's .298M_J mass.

Currently, there are no orbital simulations available for this star. A habitable planet might survive in the system as long as it was close to the outer perimeter of the habitable zone.

HD 142022A

HD 142022A is a type K0V star in the constellation Octans, The Octant, at a distance of about 116.93 light-years (RA 16 10 15.0238, D -84 13 53.802). It has a .99 $M_\odot$ mass, a 5,500±30°K effective temperature, a .19±.04 Fe/H metallicity, and 1.01$L_\odot$ luminosity. It

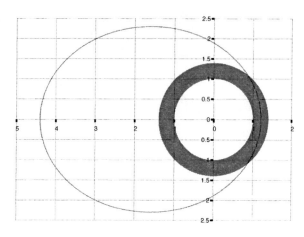

is chromospherically quiet with a rotation of 38 days. Its estimated age is 11.5 Gyrs.

This is a wide separation double-star system. Its companion star, HD 142022B, is a K7 type star, with an estimated .454 $M_\odot$ mass and .043 $L_\odot$ luminosity, is currently at a 746 AU distance. It's orbit is estimated at 22,513 years, assuming a circular orbit. An eccentric orbit could reduce this significantly. No planets have yet been found around the companion. Its age is unknown, but assumed to be the same as HD 142022A.

The giant planet, the 150th found (2005, La Silla, European Southern Observatory, Chile[76]), has a 2.8 AU semi-major axis, a 1923±80 day period, a .57 eccentricity, and a 4.4 M_J min. mass.

Currently, there are no orbital simulations available for this star. A habitable planet might survive in the system as long as it was close to the inner perimeter of the habitable zone.

GQ Lupus

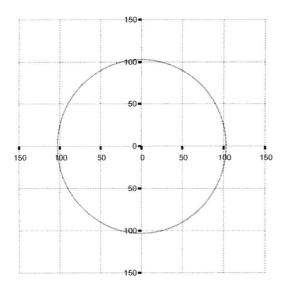

GQ Lupus is a type K7V star in the constellation Lupus, The Wolf, at a distance of about 456.4±163 light-years (RA 15 49 12.144, D -35 39 03.95). It is a young, less than 2 million year old, classical T Tauri type star in the Lupus star forming region. Based on computer models for such young stars, it has a .7 $M_\odot$ mass. The star has only recently emerged from its forming cloud and might still be accumulating mass. In addition, the internal processes of fusion are still in flux and the star has not yet stabilized into a main stream star. Thus the luminosity is uncertain and changing.

The giant planet, the 151st found (2005, Paranal, European Southern Observatory, Chile[77]), has a 103±37 AU semi-major axis, a 1,200 year period, an unknown eccentricity, and a 1–42 M_J mass. The mass of the planet, which might still be in the process of accreting material from the planetary dust disk around its primary, is very uncertain. Different planetary growth models yield different masses for the temperature of the planet, measured as being 2,000°K. This planet is one of two that were actually imaged with a telescope.

HD 13189

HD 13189 is a type K2II star in the constellation Triangulum, The Triangle, at a distance between 2,217–6,037 light-years (RA 02 09 40.1717, D +32 18 59.169). It has a 2–7 $M_\odot$ mass. Its luminosity is difficult to determine because of the uncertainty in its distance. It's an evolved giant star, and is the most massive star to have been discovered to have a planet.

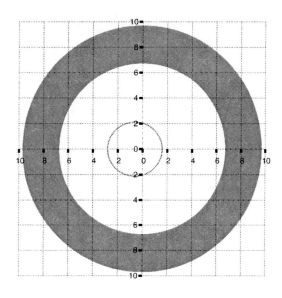

The giant planet, the 152nd found (2005, Thueringer Landessternwarte Observatory, Tautenburg, Germany, Hobby-Eberly Telescope[78]), has a 1.5–2.2 semi-major axis, a 470.9±6 day period, a .28±.06 eccentricity, and an 8–20 M_J minimum mass.

Being an evolved giant star means its habitable zone has moved from closer in to its present position. The above chart assumes the maximum distance of 2.2AU for the giant planet and a low 2.0 $M_\odot$ mass for the primary. If these are correct, a terrestrial planet at 6AU would only recently have become too hot for life. Unfortunately, the mass of the primary gives it only 740 million years on the main sequence, barely enough time for any prospective habitable planet to develop a stable surface and begin the process of generating an atmosphere.

HD 8673

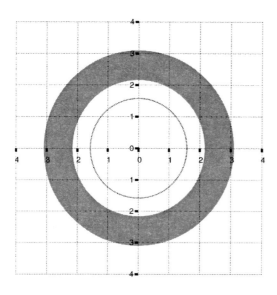

HD 8673 is a type F7V star in the constellation Andromeda, at a distance of about 124.695 light-years (RA 01 26 08.7849, D +34 34 46.921). It has a 1.3 $M_\odot$ mass. This is one of the few F-type stars with a planet.

The giant planet, the 153rd found (2005, Thueringer Landessternwarte Observatory, Tautenburg, Germany, Hobby-Eberly Telescope[78]), has a 1.58 AU semi-major axis, a 639 day period, an unknown eccentricity, and a 10–18 M_J minimum mass.

The above chart assumes the eccentricity of the planet is zero. Based on its estimated 1.3$M_\odot$ mass, this star will only spend three billion years on the main sequence. The optimal orbit for a terrestrial planet would be 2.15AU, placing it uncomfortably close to the giant planet. A path farther out from its primary, while improving the odds for a stable orbit, would also decrease the amount of time the planet spent in the habitable zone.

HD 149026

HD 149026 is a type G0 IV star in the constellation Corona Borealis, The Northern Crown, at a distance of about 257.098 light-years (RA 16 30 29.6192, D +38 20 50.315). It has a 1.3±.1 $M_\odot$ mass, a 1.45$R_\odot$ radius, a 6,147±50°K effective temperature, a .36±.05 Fe/H metallicity, and 2.72±.5$L_\odot$ luminosity.

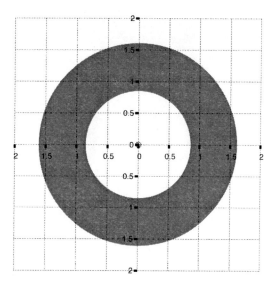

The giant planet, the 160th found (2005, Subaru Telescope, NAO Japan, Keck Observatory[100]), has a .042AU semi-major axis, a 2.8766±.001 day period, an unknown eccentricity, a .725±.05R_J radius, and a .36±.03 M_J minimum mass. This planet was found when it was observed to pass in front of its primary. Both the mass and the radius are considerably smaller than other known transiting planets. Models of close-in planets indicate a radius of 1.14R_J if it were composed of gases like our sun, and a .97R_J if it had a 20Earth-mass density (5.5g/cm³). Thus, to explain its small radius it must be composed of heavier elements. On the other hand if it were composed entirely of water-ice or olivine its radius would be .43 or .28R_J respectively. Based on computer modeling, the planet is more like Neptune, with a large heavy core, than Jupiter or Saturn.

The above chart assumes the eccentricity of the planet is zero. Based on its estimated 1.3$M_\odot$ mass, this star will only spend three billion years on the main sequence. The optimal orbit for a terrestrial planet would be 1.2AU, placing it far from the giant planet, and giving it the maximum time in the habitable zone

HD 188753

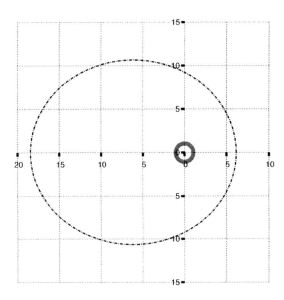

HD 188753 is a type G9 V star in the constellation Cygnus, The Swan, at a distance of about 146.3 light-years (RA 19 54 58.3705, D +41 52 17.511). It has a $1.06 M_\odot$ mass, a $1.1 R_\odot$ radius, and $1.8 L_\odot$ luminosity. Its age is unknown.

This is the first planet found in a triple-star system. The other two stars orbit the first at a distance of 12.3 AU, 25.7 year period, with a .5 eccentricity. They have a .56AU semi-major axis with a 155.5±.5 day period. They have a combined $1.6 M_\odot$ mass, with an estimated spectral type of K (orange) and M (red), respectively, approaching the other star in combined luminosity.

The giant planet, the 161st found (2005, Keck I Observatory[101]), has a .0452AU semi-major axis, a 3.35 day period, an unknown eccentricity, and a 1.14 MJ minimum mass.

The above chart places the solitary star, the G9 V, as the center of the system, with the giant planet orbiting it too close to be seen. In actual fact, the solitary star is orbiting the other two stars because they out-mass it by almost 50%.

This planet presents a problem for current planet formation theories, which assume that the giant planets form outside the "snowline" at 2.7 AU and then spiral into the close distances at which we observe them. The dual-stars, though, would disrupt and destroy any planetary formation disk around the solitary star beyond 1.3AU, rendering it impossible for any giant planets to form at the 2.7AU orbit.

No orbital simulations are available for this star, although it is highly unlikely that any stable orbits exist inside the habitable zone.

Appendix A: Withdrawn Discoveries

HD 219542B

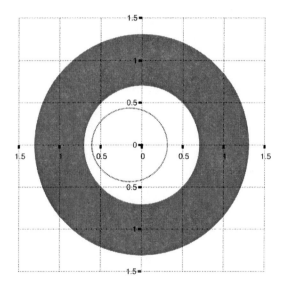

HD 219542B (HIP 114914, BD -02°5917, ADS 16642) is one of a wide pair of main sequence type G2V and G7V stars in the constellation Pisces, The Fishes, at a distance of about 160–199 lightyears (RA 23 16 35.402, D -01 35 08.40). HD 219542A, type G2V, has an 8.19 apparent magnitude, B is a type G7V with an 8.58 apparent magnitude. Based on apparent magnitude and distance, A has a $1.13\pm.05M_\odot$ mass, a $1.05R_\odot$ radius, an effective 5,989°K temperature, a .29 FeH ratio, and a rotation of 26.6 days. B has a $1.06\pm.05M_\odot$ mass, a $.96R_\odot$ radius, an effective 5,732±31°K temperature, a .17±.04 FeH ratio, and a rotation of 21 days. Oddly, A has an estimated age of 5.7 Gyrs while B is estimated at 2.7 Gyrs. Both exhibit low chromospheric activity. These two stars are separated by about 288AU, however their orbital eccentricity is unknown.

Like 16 Cyg B and HD 178911B, the proposed giant planet is orbiting the lithium poor component of the system, HD 219542B.

The giant planet, which would have been the 103rd found (Osservatorio Astronomico di Padova, Italy[83]), had a .46AU semi-major axis, a 112.1±1.2 day period, a .32±.15 eccentricity, and a .30M_J minimum (approximately the same as Saturn). The founders had a 96 percent confidence level that this was a planet, especially because they had simultaneously examined its near twin companion and seen no evidence of any giant planets. Nothing in their data excluded the possibility of terrestrial planets in either system.

The second paper[84] released by the discoverers, however, withdraws the proposed planet discovery. Further observations have led them to conclude the radial velocity measurements indicating a planet are instead artifacts of a temporary, about one year, burst of activity in the star's chromosphere.

Appendix B: Tentative Discoveries

Transit Detection

One method of planet detection is by transit of the planet across the face of its primary. Basically, the searcher takes a series of light measurements of the target star. If a planet passes in front of the star during these observations it causes a "dip" in the light output measured. If the orbit of the planet is close enough, several such passages can be detected during the observation time. The percentage of light blocked by the planet, as well as the steepness of the dip in the light curve, yield a fairly accurate estimate of the size of the transiting planet.

Orbital eccentricity cannot be determined from these measurements. If the spectral type of the star is unknown then the mass of the planet cannot be determined. All that is known is the orbital period and planet radius. Followup observations using radial velocity measurements are required to determine these details.

Currently, most of these transits are being detected by the MACHO Project. This project's aim is to test the hypothesis that a significant fraction of the dark matter in the halo of the Milky Way is made up of objects like brown dwarfs or planets: these objects have come to be known as MACHOs, for MAssive Compact Halo Objects.

The latest paper, which examined 180,000 brightest stars in fourteen 0.5 deg^2 fields toward the Galactic Bulge, yielded several hundred transits. Using the shapes and colors of these objects to separate the transits of objects with relatively small radii from the grazing eclipses of stars with larger radii, nine objects with short-period eclipses consistent with either giant extrasolar planets or late-type dwarf stars were found.

All nine are over 3,000 parsecs away (9,780 light-years):

Name	Radius	Orbital Period	Constellation
MACHO-104.20644.308	$2.3R_J$	2.28316 days	Sagitarius
MACHO-118.18272.189	$1.7R_J$	1.96730 days	Sagitarius
MACHO-118.18407.57	$2.4R_J$	2.39860 days	Sagitarius
MACHO-118.18793.469	$2.7R_J$	2.03720 days	Sagitarius
MACHO-120.22041.3265	$2.8R_J$	5.40830 days	Sagitarius
MACHO-120.22303.5389	$1.8R_J$	2.43365 days	Sagitarius
MACHO-130.45607.284	$2.2R_J$	1.69966 days	Scutum
MACHO-402.47741.740	$3R_J$	3.71120 days	Sagitarius
MACHO-402.47800.723	$2.9R_J$	4.27480 days	Sagitarius

Optical Gravitational Lensing Experiment

A second method of planet detection is OGLE, the Optical Gravitational Lensing Experiment. This process, also known as microlensing, uses the fact that gravity bends light to find "dark" objects. The idea is if another object passes directly between an observer and a star, the gravity of the intervening object will cause the image of the star to split into two images and then recombine. The separation distance of the images, the time length of the splitting, and the speed at which the split develops all give important information about the object that passed between the observer and the star.

The Magellanic Clouds and the Galactic Bulge are the most natural locations to conduct such searches due to the many of background stars that are potential targets for microlensing. The LMC and SMC stars may be lensed mostly by the Galactic halo objects. In the case of the Galactic Bulge stars one expects an additional component — microlensing by low-mass disk stars.

While the OGLE project has detected five planets, four of these were by transits. Only the fifth is an actual microlensing event: OGLE 2003-BLG-235/MOA 2003-BLG-53. This event was really a double event: an observed star was seen to split into three images, collapse into two images of unequal size, then back into one. The conclusion was that both a planet and the star it is orbiting passed in front of the observed star.

Depending on the lens model chosen, the planet has either a $1.5M_J$ or a $2.5M_J$ mass, and is located 5,200 parsecs (16,952 light-years) away in the constellation Sagitarius. Nothing is known about the parent star.

One of the earliest microlensing detections was accomplished by MACHO-97-BLG-41 in 1997[93]. Analysis of that event gives the conclusion that a $3.5\pm1.8M_J$ mass planet is orbiting a double-star system at a distance of about 7AU. The stars are a late K dwarf ($.6\pm.3M_\odot$) and an M dwarf ($.16\pm.08M_\odot$) separated by 1.8AU. Distance to the system is estimated at $6.3^{+.6}_{-1.3}$ kpc ($20,538^{+1,956}_{-4,238}$ light-years).

And winning the contest for the most distant planet discovered is Q0957+561 in Ursa Major, at 2.4 Giga light-years. It was discovered by accident while observing the unusual Quasar Q0957+561, which is split into two images in the sky by an invisible (not bright enough to be seen) galaxy in front of it. The two images are different sizes and the smaller image is 1.1 years older than the other image. The scale and duration of a tiny fluctuation in the image indicates a planetary object of one to three times the size of Earth passed through our line of sight.

Dust Rings

The third method of detecting planets is very indirect. Several papers have detailed gaps in the dust rings surrounding certain young stars. The postulate is that these gaps are caused by the gravitational interference of a planet, much as gravity of the moons of Saturn has created gaps in its rings.

GM Aurigae, a young T Tauri (a star between 100,000 and 1 million years old with erratic luminosity) is a K5Ve type star, located in the constellation Auriga, The Charioteer. It appears to have a 2 M_J mass planet,[94] orbiting at 2.5AU in a disc with 0.047 $M_\odot$ mass and 300 AU radius, and provides a good match to the observations which constrain the velocity field in the disc. A range of planet masses is allowed by current data and more observations are needed.

TW Hydrae 6 (HD 53911) is another T Tauri, K5Ve type star in the constellation Hydra, The Water Serpent, about 183.9 light-years away. Careful examination has revealed a very red point source at a distance of 125AU. It appears to be a 2M_J mass planet. There is also a clearing of dust from the inner area of the star, similar to that of GM Aurigae. It is proposed that this clearing might be evidence of a planet. Examinations of the companion stars in the group has revealed that two of them have a brown dwarf companion, one at 100 AU with a 20M_J mass and the other has one at 180AU with a 40M_J mass.

Fomalhaut (HD 216956, HR 8728), the seventeenth brightest star in the sky, in the constellation Piscis Austrinus, the Southern Fish, is an A3V type star about 25 light-years away (RA 22 57 39.1, Dec -29 37 20). The system is only about 200 million years old and has a dust ring that looks somewhat like a donut seen slightly edge-on. The empty hole in the middle seems indicative of the presence of a Saturn-sized planet that has swept up the dust in that area.

Free-floating Planets

S Orionis 70. Located in the constellation Orion is a dense gas starforming region. In this nebula is a small object (RA = 05 38 10.1, D -02 36 26) with an estimated $3^{+2.5}_{-1.5} M_J$ mass, a .16R solar radius, a $1,100^{+200}_{-100}$ °K effective temperature, 440pc/1,434 lys distant, and an estimated age of 3 million years.[91] It appears to be a planet that has been ejected from one of the systems in the Orionis cluster. This is the only case known of a solitary, ejected planet. Its position in the nebulae appears to indicate that such ejections are common events.

However, another group[92] has examined the object and concluded it is instead a massive, much older brown dwarf star in front of the Orionis cluster.

Appendix C: The Constellations

The star symbol ☆ indicates the original 48 constellation.

Some constellations observable in the sky at night are actually smaller pieces of a larger constellation. The Big Dipper, for instance, is actually the center portion of the constellation Ursa Major.

(a) Indicates the 12 constellations of the zodiac, that band of constellations that lie along the ecliptic and are in the apparent annual path of the sun and monthly path of the moon.

(b) Carina, Puppis, and Vela are subdivisions of the original constellation Argo Navis, the legendary ship Argo. It was broken up to make for a more manageable arrangement. Some sources list the constellation Pyxis as a part of it.

(c) Contains two parts separated by the constellation Ophiuchus. Serpens Cauda, the head, is the eastern half. Serpens Caput, the tail, is the western half. It is considered one constellation, anyway.

	Latin Name	*Genitive*	*Abbreviation*	*English Name*
☆	Andromeda	Andromedae	And	Andromeda
	Antila	Antliae	Ant	The Air Pump
	Apus	Apodis	Aps	The Bird of Paradise
☆	Aquarius (a)	Aquarii	Aqr	The Water Carrier
☆	Aquila	Aquilae	Aql	The Eagle
☆	Ara	Arae	Ara	The Altar
☆	Aries (a)	Arietis	Ari	The Ram
☆	Auriga	Aurigae	Aur	The Charioteer
☆	Bootes	Bootis	Boo	The Herdsman
	Caelum	Caeli	Cae	The Graving Tool
	Camelopardalis	Camelopardalis	Cam	The Giraffe
☆	Cancer (a)	Cancri	Cnc	The Crab
	Canes Venatici	Canum Venaticorum	CVn	The Hunting Dogs
☆	Canis Major	Canis Majoris	CMa	The Larger Dog
☆	Canis Minor	Canis Minoris	CMi	The Smaller Dog

Latin Name	Genitive	Abbreviation	English Name
☆ Capricornus (a)	Capricorni	Cap	The Goat
☆ Carina (b)	Carinae	Car	The Keel (of the ship)
☆ Cassiopeia	Cassiopaeiae	Cas	Cassiopeia
☆ Centaurus	Centauri	Cen	The Centaur
☆ Cephus	Cephei	Cep	Cepheus
☆ Cetus	Ceti	Cet	The Whale
Chamaeleon	Chamaeleontis	Cha	The Chameleon
Circinus	Circini	Cir	The Pair of Compasses
Columba	Columbae	Col	The Dove
Coma Berenices	Comae Berenices	Com	Berenice's Hair
☆ Corona Australis	Coronae Australis	CrA	The Southern Crown
☆ Corona Borealis	Coronae Borealis	CrB	The Northern Crown
☆ Corvus	Corvi	Crv	The Crow
☆ Crater	Crateris	Crt	The Cup
Crux	Crucis	Cru	The Cross
☆ Cygnus	Cygni	Cyg	The Swan
☆ Delphinus	Delphini	Del	The Dolphin
Dorado	Doradus	Dor	The Swordfish
☆ Draco	Draconis	Dra	The Dragon
☆ Equuleus	Equulei	Equ	The Little Horse
☆ Eridanis	Eridani	Eri	The River
Fornax	Fornacis	For	the Furnace
☆ Gemini (a)	Geminorum	Gem	The Twins
Grus	Gruis	Gru	The Crane
☆ Hercules	Herculis	Her	Hercules
Horologium	Horologii	Hor	The Clock
☆ Hydra	Hydrae	Hya	The Water Serpent
Hydrus	Hydri	Hyi	The Water Snake
Indus	Indi	Ind	The Indian
Lacerta	Lacertae	Lac	The Lizard
☆ Leo (a)	Leonis	Leo	The Lion
Leo Minor	Leonis Minoris	LMi	The Smaller Lion
☆ Lepus	Leporis	Lep	The Hare
☆ Libra (a)	Librae	Lib	The Scales
☆ Lupus	Lupi	Lup	The Wolf
Lynx	Lyncis	Lyn	The Lynx
☆ Lyra	Lyrae	Lyr	The Lyre
Mensa	Mensae	Men	The Table Mountain
Microscopium	Microscopii	Mic	The Microscope
Monoceros	Monocerotis	Mon	The Unicorn
Musca	Muscae	Mus	The Fly
Norma	Normae	Nor	The Ruler
Octans	Octantis	Oct	The Octant
☆ Ophiuchus	Ophiuchi	Oph	The Serpent Carrier
☆ Orion	Orionis	Ori	Orion
Pavo	Pavonis	Pav	The Peacock
☆ Pegasus	Pegasi	Peg	Pegasus

The Constellations

Latin Name	Genitive	Abbreviation	English Name
☆ Perseus	Persei	Per	Perseus
Phoenix	Phoenicis	Phe	The Phoenix
Pictor	Pictoris	Pic	The Easel
☆ Pisces (a)	Piscium	Psc	The Fishes
☆ Piscis Austrinus	Piscis Austrini	PsA	The Southern Fish
☆ Puppis (b)	Puppis	Pup	The Stern (of the ship)
Pyxis (b)	Pyxidis	Pyx	The Mariner's Compass
Reticulum	Reticuli	Ret	The Net
☆ Sagitta	Sagittae	Sge	The Arrow
☆ Sagittarius (a)	Sagittarii	Sgr	The Archer
☆ Scorpius (a)	Scorpii	Sco	The Scorpion
Sculptor	Sculptoris	Scl	The Sculptor's Tools
Scutum	Scuti	Sct	The Shield
☆ Serpens (c)	Serpentis	Ser	The Serpent
Sextans	Sextantis	Sex	The Sextant
☆ Taurus (a)	Tauri	Tau	The Bull
Telescopium	Telescopii	Tel	The Telescope
☆ Triangulum	Trianguli	Tri	The Triangle
Triangulum Australe	Trianguli Australis	TrA	The Southern Triangle
Tucana	Tucanae	Tuc	The Toucan
☆ Ursa Major	Ursae Majoris	UMa	The Larger Bear
☆ Ursa Minor	Ursae Minoris	UMi	The Smaller Bear
☆ Vela (b)	Velorum	Vel	The Sails (of the ship)
☆ Virgo (a)	Virginis	Vir	The Virgin
Volans	Volantis	Vol	The Flying Fish
Vulpecula	Vulpeculae	Vul	The Fox

Appendix D: Masses and Orbital Characteristics of Extrasolar Planets

R.A = Right Ascension; Dec = Declination; Star = Star's mass where Sol = 1; Planet = Planet's mass where Jupiter = 1; Period = time to complete one orbit; Axis = Semi-major axis distance in Astronomical Units (92,960,116 miles); Luminosity = star brightness where Sol = 1. Data from SIMBAD, California & Carnegie Planet Search (http://exoplanets.org/planet_table.shtml), Extra-solar Planets Catalog (http://www.obspm.fr/planets), and Statistical Properties of Exoplanets (N. Santos et al., submitted 11/11/02).

Luminosity Range: On those stars with a range given for luminosity, the first number is the luminosity in ultraviolet and the second number is the luminosity in the visible light range of yellow-green. Typically, if the first number is higher it is a "hot" star, if the second number is higher, it is a "cool" star.

#	Star Name HD	Hip	Common	R.A.	Dec.	Distance Parsecs/lyrs	Type	Star Sol=1	Planet Jup=1	Period days	Axis (AU)	Orbit ecc.	Lum Sol=1
1	114762	64426	BD +18 2700	13 12 19.7427	+17 31 01.643	28/91.28	F9V	.80	9–11.02	84.03±.134	0.351	0.33±.01	5.61–5.06
2	217014	113357	51 Peg	22 57 27.14	+20 46 04.5	14.7/47.922	G2IV	1.04	0.46±.02	4.231	0.052±.001	0.01±.003	.63–1.17
3	117176	65751	70 Vir	13 28 26.54	+13 47 12.43	22/71.72	G4V	.92±.18	7.41±.8	116.7	0.482±.05	0.40±.01	2.25–3.77
4	95128b	53721b	47 UMa b	10 59 27.9737	+40 25 48.925	14.08/45.90	G0V	1.05±.03	2.56±.02	1,090±3	2.09	0.06±.014	.92–1.54
67	95128c	53721c	47 UMa c						0.76	2,640±90	3.78±.05	0.00±.01	
5	75732b	43587b	55 Cnc b	08 52 37.60	+28 20 02.6	13.4/43.68	G8V	.88±.15	.784±.09	14.67±.001	0.115±.003	0.0197±.012	.287–.64
78	75732c	43587c	55 Cnc c					.217±.04	43.93 ±.25	0.24±.008	.44±.08		
79	75732d	43587d	55 Cnc d					4,517.4±77.8	5.257±0.208	327±.28			
127	75732	43587e	55 Cnc e	.045±.01	2.808±.002	.038±.001	3.912±.52 .174±.127						
6	120136	67275	Tau Boo	13 47 17.345	+17 27 22.31	15/48.9	F7V	1.3	4.14±.06	3.313±.001	0.047±.03	0.04±.01	1.96–3.04
7	95735b	54035b	Lalande 21185	00 03 20.1939	+35 5811.547	2.5/8.25	M2V	.46	.9	5.8 yrs	2	low	.006
8	95735c	54035c	Lalande 21185c	1.6	30yrs	10	low						
9	9826b	7513b	Ups And b	01 36 48.527	+41 24 38.71	13.47/43.9	F8V,	1.29	0.68±.01	4.617±.0003	0.059	0.012±.15	2.18–3.58
21	9826c	7513c	Ups And c					1.94±.05	241.3±1.2	0.828±.001	0.25±.13	0.03±.01	
22	9826d	7513d	Ups And d					4.02±.27	1.299±30	2.54±.01	0.31±.11		
10	186427	96901	16 Cyg b	19 41 51.9720	+50 31 03.083	21.4/69.76	G3V	.99±.02	1.68±.18 (14 – Han Et al 2000)	798.4±6	1.69±.03	0.68±.01	.70–1.29
11	143761	78459	Rho Crb	16 01 02.66	+33 18 12.634	16.7/54.44	G0Va	1.0±.05	0.99±.08	39.81±.24	0.224±.06	0.07±.15	.94–1.65
12	Ross 780b	113020b	Gl876b	22 53 16.7339	-14 15 49.322	4.69/15.32	M4V	.32	1.935±. .007	60.94	0.207	0.0249	.002
55	Ross 780c	113020c	Gl876c						0.56	30.12	0.130	0.27	
154	Ross 780c	113020d	Gl876d						.023±.003	1.93776	.0208067	0.0	
13	145675	79248	14 Hercules	16 10 24.3143	+43 49 03.521	18.1/59.2	K0V	.79	3.3	4.52yrs	2.5	.326	.63
14	187123	97336		19 46 58.1130	+34 25 10.288	50/163	G5V,	1.0±.05	0.54±.03	3.097	0.042	0.03±.01	.875–1.53
15	210277	109378		22 09 29.49	-07 32 32.7	22/71.72	G0/G7	.99±.05	1.29±.05 (poss. 2.2±.6)	436.6±1.5	1.12 ±.11	0.45±.051	.48–.92
16	195019	100970		20 28 18.6363	+18 46 10.188	37.36/121.79	G3IV-V	1.02±.04	3.55±.05	18.20	0.136±.004	0.02±.02	2.07–4.02
17	217107b	113421b	HR 8734b	22 58 15.45	-02 23 42.4	19.72/64.25	G8IV	.98	1.37±.14	7.1269±.00022	0.074	0.13±.02	.577–1.12
158	217107c	113421c	HR 8734c						2.1±.1	3,150±1,000	4.3±2	.55±.2	
18	13445	10138	GJ 86	02 10 14.42	-50 50 00.5	11/35.86	K1V	.79	4.00±.02	15.78±.02	0.11±.007	0.04	.17–.35
19	168443b	89844b	GJ 4052b	18 20 04.11	-09 35 34.6	33/107.58	G8IV/V	1.01±.05	7.64±.44	58.10	0.295	0.53±.02	.83–1.59
	168443c	89844c	GJ 4052c						16.96±.027	1,770±.25	2.87	0.20±.02	
20	75289	43177		08 47 40.3894	-41 44 12.452	28.94/94.344	G0V	1.05	0.42±.01	3.50±.009	0.047±.003	0.00±.05	1.2–2.0
23	17051	12653	Iota Hor. –HR 810	02 42 31.65	-50 48 12.3	15.5/50.53	G0V	?M	2.25±.18	312.0±10	0.909±.104	0.15±.08	.84–1.42
24	130322	72339		14 47 32.73	-00 16 53.3	30/97.8	K0V	.79±.25	1.15±.07	10.72	0.092±.04	0.05±.002	.23–.46

		Star Name		R.A.	Dec.	Distance	Type	Star	Planet	Period	Axis	Orbit	Lum
25	192263	99711		20 13 59.8451	-00 52 00.757	19.9/64.87	K0	.79±.04	.75	23.87±.14	.15	.03±.16	.13-.19
26	209458	1008859		22 03 10.7	+18 53 04	47/153.22	G0V	1.05±.1	0.63±.06	3.52433±.00027	0.046±.01	0.02±.02	1.01-1.64
27	10697	8159		01 44 55.8246	+20 04 59.338	30/97.8	G5IV	1.1±.12	6.08±.03	1,074±7	2.12	0.11±.02	1.26-2.34
28	37124b	26381b		05 37 02.4864	+20 43 50.836	33/107.58	G4V	.91±.15	0.61±.04	152.46±.369	0.53±.003	0.055	.42-.79
81	37124c	26381c							.6	843.6	1.64	0.14	
159	37124d	26381d							.66	2,295	3.19	.2	
29	134987	74500		15 13 28.6676	-25 18 33.649	25/81.5	G5	1.05	1.63±.02	265±5.3	0.821±.01	0.37±.12	.73-1.4
30	177830	93746		19 05 20.7735	+25 55 14.379	59/192.34	K0	1.17±.14	1.24±.24	391±18	1.10 ±.1	0.40±31	1.46-4.01
31	222582	116906		23 41 51.5299	-05 59 08.726	42/136.92	G5	1.0±.02	5.20±.02	577.1±1.4	1.36	0.76±.01	.72-1.25
32	1237	1292	GJ3021	00 16 12.6775	-79 51 04.254	17.62/57.44	G6V	0.9	3.45±.13	133.8±.02	0.505±.018	0.51±.02	.35-.55
33	89744	50786	HR 4067	10 22 10.5621	+41 13 46.308	40/130.4	F7V	1.40	7.17±.57	256.0±.734	0.883±.007	0.70±.02	4.39-6.91
34	12661b	9683b		02 04 34.29	+25 24 51.5	37.16/121.14	G6V	1.07±.02	2.30	263.3±1.2	0.823	0.35±.03	.05-.14
80	12661c	9683c							1.56±.01	1,444.5±12.5	2.56	0.20±.04	
35	16141	12048		02 35 19.9283	-03 33 38.167	35.9/117.03	G5IV	1.0	0.22±.01	75.80±.4	0.35	0.21±.7	1.11-2.13
36	46375	31246		06 33 12.6237	+05 27 46.532	33.4/108.88	K1IV	1.0±.17	0.25±.01	3.024	0.04±.001	0.02±.02	.31-.69
37	52265	33719b	HR 2622	07 00 18.0363	-05 22 01.783	28/91.28	G0V	1.13	1.00±.19	119.0±.1	0.49±.008	0.29±.09	1.23-2.03
38		1120404	BD-10 3166	10 58 28.780	-10 46 13.39	?	K0V	1.10	0.48	3.488±1	0.046	0.06	
39	82943b	47007b		09 34 50.7361	-12 07 46.365	27.46/89.52	G0	1.05±.1	1.63	444.6±8.8	1.16	0.41±.08	.90-1.56
62	82943c	47007c							0.88	221.6±2.7	0.73±.002	0.54±.05	
40	83443b	47202b		09 37 11.8281	-43 16 19.939	43.54/141.90	K0V	.79	0.34±.01	2.985	0.038	0.079± .008	.39-.82
48	83443c	47202c							.15	29.83	.17	.42	
41	108147	60644		12 25 46.2686	-64 01 19.516	38.57/125.73	F8/G0V	1.27±.31	0.40±.01	10.9 ±.001	0.079±.021	.48±.09	2.03
42	168746	90004		18 21 49.7832	-11 55 21.660	43.12/140.57	G5	.92	0.24±.01	6.400±.004	0.066±.01	.081±.081	1.05
43	169830b	90485b		18 27 49.4838	-29 49 00.715	36.32/118.40	F8V	1.4	2.95±.01	230.4±4.5	0.823	0.34±.05	3.14-4.87
114	169830c	90485c							4.04	2,102±264	3.6	.33±.02	
	162020	87330		17 50 38.3575	-40 19 06.056	31.26/101.90	K2V	.7	13.73±.02	8.42	0.072	0.28±.013	.17
44	38529b	27253b		05 46 34.91	+01 10 05.496	42.43/138.32	G4IV	1.39	0.78	14.31±.05	0.129	0.28	3.26-6.49
82	38529c	27253c							12.78±.08	2,207.4±33	3.71±.03	0.33	
45	92788	52409		10 42 48.5287	-02 11 01.521	32.82/106.99	G5	1.06±.04	3.59±.29	337±40	0.969±.019	0.28±.06	.84-1.33
46	6434	5054		01 04 40.1511	-39 29 17.583	40.32/131.44	G2/3V	1.0	0.48	22.09	0.154±.004	0.30±.05	1.13
47	19994	14954	94 Cet	03 12 46.4365	-01 11 45.964	22.38/72.95	F8V	1.35±.01	1.66±.34	454±19	1.19±.11	0.20	2.39-4.0
49	121504	68162		13 57 17.2375	-56 02 24.153	44.37/144.64	G2V	1.0	0.89	64.62±.07	0.317±.003	0.13±.06	.94-1.62
50	190228	98714		20 03 00.7730	+28 18 24.685	66.11/215.51	G5IV	1.3 ±.19	3.44-5.0	1,112±42	1.98±3.3	0.43±.08	2.23-4.49
51	22049	16537b	Eps Eri	03 32 55.8442	-09 27 29.744	3.2p/10.43	K2V	0.8	0.92±.06	2,550±48	3.39±.07	0.43	.12-.28
									0.1?	>140yrs	40?	.3?	

#	Star	Name	R.A.	Dec.	Distance	Type	Star	Planet	Period	Axis	Orbit	Lum
52	179949	94645	19 15 33.2278	-24 10 45.668	27/88.02	F8V	1.24±.01	0.93±.05	3,092±.001	0.045±.005	0.00±.03	1.23–1.96
53	27442	19921	04 16 29.0287	-59 18 07.763	18.1/59.00	K2IVa	1.2	1.32±.11	415±146	1.16±.02	0.06±.04	1.74–4.69
54	160691b	86796b	HR 1355	17 44 08.7029	-51 50 02.591	15.3/49.87	G3IV-V	1.08±.02	1.68±0.2	638±10	1.48±0.1	0.2±.08	.91–1.74
125	160691c	86796c	Mu Arae b						14ME	9.55±.03	.09	0.0±.02	
135	160691d	86796d	Mu Arae c						3.1	8.2yrs	4.16	.57	
56	8574	6643	Mu Arae d	1 25 12.5168	28 34 00.096	44.15/143.93	F8	1.17	2.08±.17	228.18±.68	0.77±.01	0.31±.09	1.52–1.26
57	28185	20723		04 26 26.3205	-10 33 02.955	39.4/128.44	G5	0.99	5.70	383±2	1.03	0.07±.04	.51–.99
58	50554	33212		06 54 42.8253	+24 14 44.011	31.03/101.15	F8	1.1±.04	3.72±1.8	1,254±25	2.32±.06	0.51±.01	.909–1.48
59	74156b	42723b		08 42 25.1222	+04 34 41.151	64.56/210.46	G0	1.05±.22	1.55±.01	51.60±.053	0.276	0.65±.022	1.94–3.19
60	74156c	42723c							7.46±.04	2300.0	3.47	0.39±.074	
61	80606	45982		09 22 37.5679	+50 36 13.397	58.38/190.31	G5	0.9±.13	3.43±.02	111.78±.21	0.438±.001	0.93±.012	.40–.78
63	106252	59610		12 13 29.5093	+10 02 29.898	37.44/122.05	G0	1.05±.03	6.79±.02	1,503±1.5	2.53±.01	0.57	.75–1.36
64	141937	77740		15 52 17.5474	-18 26 09.834	33.46/109.08	G2/G3V	1.0±.1	9.67±.03	658±6	1.48±.04	0.40±.01	.69–1.20
65	178911	94705		19 09 03.1039	+34 35 59.45	46.73/152.33	G5	.87	6.46±.17	71.50±.013	0.32±.06	0.14±.016	.61–1.20
66	213240	111143		22 31 00.3672	-49 25 59.773	40.75/132.84	G0/G1V	1.22	4.49	951.0	2.02	0.45	1.54–2.70
68	4203	3502		00 44 41.2021	+20 26 56.138	77.5/252.65	G5	1.06±.13	1.64	406.0±3	1.09	0.53±.02	.88–1.72
69	4208	3479		00 44 26.6503	-26 30 56.449	33.9/110.51	G5V	0.93±.07	0.81	829.0±2	1.69	0.04	.40–.75
70	33636	24205		05 11 46.4490	+04 24 12.742	28.7/93.5	G0V	.99±.13	7.71±.11	1,553±67	2.62±.2	0.39±.02	.62–1.05
71	68988	40687		08 18 22.1731	+61 27 38.599	58/189.08	G0	1.2	1.90	6.276	0.07±.001	0.14	.84–1.49
72	114783	64467		13 12 43.7860	-02 15 54.143	22/71.72	K0	.92±.04	0.99±.02	501±1.14	1.20	0.10±.01	.16–.38
73	142	522		00 06 19.1755	-49 04 30.688	25.6/83.5	G1 IV	1.1±.16	1.14±.14	331.8±7.1	0.980	0.37±.01	1.18–1.9
74	23079	17096		03 39 43.0952	-52 54 57.017	34.8/113.4	F8/G0V	1.1±.1	2.76±.23	628±1	1.48±.01	0.14±.1	.94–1.49
75	39091	26394	Pi Men	05 37 09.8917	-80 28 08.839	20.55/66.99	G1 IV	1.1	10.39±.04	2,280±212	3.50±.21	0.63±.01	1.14–1.94
76	137759	75458	Iota Draconis	15 24 55.7747	+58 57 57.836	31.5/102.69	K2III	1.05	8.68±.04	550±.6	1.34	0.71	3.53–40.61
77	136118	74948		15 18 55.4719	-01 35 32.590	52.3/170.5	F9V	1.24±.04	11.91±.2	1,209±24	2.39±.09	0.37±.025	2.5–4.06
83	49674	32916		06 51 30.5164	+40 52 03.923	40.7/132.58	G5V	1.0	0.12	4.948±.001	0.057±.003	0.00±.18	.425–.814
84	72659	42030		08 34 03.1895	-01 34 05.583	51.4/167.56	G0V	.95	2.54±.01	2185	3.24	0.18	.359–2.298
85	108874b	61028b		12 30 26.8829	22 52 47.383	68.5/223.31	G5V	1.0±.04	1.36±.13	395.4±2.5	1.05±.02	0.07±.04	.654–1.25
155	108874c	61028c							1.018±.3	1,605.8±88	2.68±.25	.25±.07	
86	114729	64459		13 12 44.2575	-31 52 24.056	35/114.1	G0V	.93±.01	0.88±.02	1,136±2.5	2.08	0.33±.01	1.24–2.20
87	128311b	71395b		14 36 00.5607	.09 44 47.466	16.6/54.11	K3V	.80	2.58	420.514	1.02	0.3	.093–.233
156	128311c	71395c							3.21±.3	919	1.76±.13	.17±.09	
88	30177	21850		04 41 54.3731	-58 01 14.725	55/179.3	G8V	.95±.05	7.64±.05	1,620±2	2.65±.01	0.21±.01	.555–1.12
89	73526	42282		08 37 16.4839	-41 19 08.767	99/322.74	G6V	1.02	3.63±.75	188.0±2	0.647±.013	0.52±.21	1.01–2.10
90	196050	101806		20 37 51.7102	-60 38 04.135	46.9/152.89	G3V	1.1	3.0±0.5	1,288±230	2.4±0.5	0.28±.15	1.08–1.71

	Star Name		R.A.	Dec.	Distance	Type	Star	Planet	Period	Axis	Orbit	Lum
91	216437	113137	22 54 39.4833	−70 04 25.352	26.5/86.39	G4 IV-V	1.07±.08	2.1±0.3	1,294±250	2.3 ±0.5	0.33±.09	1.26–2.25
	13507	10321	02 12 55.0057	+40 40 06.023	26.2/85.412	G4	.93	3.19	1,318.0	2.30	0.13±.09	.421–.781
withdrawn 9/10/02												
92	20367	15323	03 17 40.0461	31 07 37.372	27.–1/88.02	G0	1.17	1.12±.05	500±9.3	1.28±.03	0.32±.09	1.06–1.70
93	23596	17747	03 48 00.3739	40 31 50.287	52.0/169.52	F8	1.29±.01	8.0±.81	1,548±25	2.87±.01	0.298±.028	.705–2.935
94	150706	80902	16 31 17.5856	79 47 23.189	27.2/88.67	G0	.98±.23	1.0	264.9±5.8	0.82	0.38±.12	.58–.978
95	190360b	98767b	20 03 37.4055	29 53 48.50	15.9/51.834	G6IV	.9±.06	1.502±.13	2,891±85	3.92±.2	0.48	.57–1.12
157	190360c	98767c						.057±.015	17.1±.015	.128±.002	.01	
96	114386	64295	13 10 39.8231	−35 03 17.218	28/91.28	K3V	.68	0.99	872±34	1.62	0.28±.1	.089–.320
97	147513	80337	16 24 01.2899	−39 11 34.729	12.9/42.05	G3/5V	.92±.19	1.00	540.4±4.4	1.26	0.52±.08	.578–1.002
	223084	117258	23 46 34.1141	−08 59 48.738	38.58/125.77	G0	?	1.21	104.1	0.44	0.48	1.02–1.67
withdrawn 9/10/02												
98	2039	1931	00 24 20.2778	−56 39 00.171	89.8/292.74	G2/3 IV/V	.98±.22	5.1±.27	1,190.0±7	2.2±.02	0.69	.98–1.71
99	76700	43686	08 53 55.5153	−66 48 03.571	59.7/194.62	G6 V	1±.05	0.19±.017	3,971±.001	0.049±.004	0.00±.04	.85–1.76
100	216435	113044	22 53 37.9315	−48 35 53.828	33.3/108.55	G0 V	1.25±.01	1.23	1326	2.6	.14	2.05–3.67
101	222404	116727	23 39 20.8490	+77 37 56.193	11.8/38.46	KIIV	?	1.76	903±6	2.1	.2±.07	2.36–6.10
102	40979	28767	06 04 29.9431	44 15 37.599	33.3/108.55	F8 V	1.08	3.16±.16	260±7	0.818 ±.07	0.26±.03	1.17–1.903
103			17 56 35.51	−29 32 21.2	1500/5000	G2V	1.04	1.45±.23	29 hours	.0225	0.0	1?
104	3651	3093	00 39 21.8061	+21 15 01.701	11/36.23	K0V	.79	.20	62.23	.284	0.0	.21–.505
105	47536	31688	−32 20 23.045	123/395	K0III	1.3–3.0	4.96–9.67	712	1.61–2.25	.20	.63	
106			17 51 48.95	30 13 25.1	3000/9780	F9/G05	1.0	.5	1.1899	.043	33–99	1?
107	73256	42214	08 36 23.0155	−30 02 15.456	36.5/118.99	G8/K0	1.05	1.85	2.54863	.037	.038	.335–.69
108	104985	58952	12 05 15.1178	+76 54 20.641	102/332.52	G9III	1.5	6.3	198.2±.3	.78	.03±.02	19–42
109	10647	7978	01 42 29.3157	−53 44 27.003	17.4/56.724	F8V	1.07	.91	1,040±37	2.1	.18±.08	.979–1.59
110	41004A	28393	05 59 49.6493	−48 14 22.889	43/140.18	K1V	.7	2.3	655±37	1.31	.39±.17	.253–.65
111	65216	38558	07 53 41.3223	−63 38 50.363	34.3/111.818	G5V	.92	1.21	613.1±11.4	1.37	.41±.06	.386–.696
112	111232	62534	12 48 51.7543	−68 25 30.544	29/94.54	G8V	.78	6.8	1,143±14	1.97	.20±.01	.334–.664
113	142415	78169	15 57 40.7907	−60 12 00.926	34.2/111.492	G1V	1.03	1.62	386±1.6	1.05	.5	.681–1.18
114	216770	113238	22 55 53.7097	−26 39 31.547	38/12.88	K1V	.90	.65	118.45±.4	.46	.37	.332–.707
115	70642	40952	08 21 28.1361	−39 42 19.474	28.8/93.79	G5IV-V	1.0±.05	2.0	2,231±400	3.3	.10±.06	.494–.950
116	330075	77517	15 49 37.6913	−49 57 48.692	50.20/163.652	G5	.95	.76	3.369±.004	.043	0.0	5.7–8.12
117	59686	36616	07 31 48.3969	+17 05 09.765	92/299	K2III	2.6	6.5	303	0.8	?	16–48
118	219449	114855	23 15 53.4947	−09 05 15.853	45/146.7	K0III	2.4	2.9	182	0.3	?	13–35.8
	54719	34693	07 11 08.3707	+30 14 42.587	92.5/301	K2III	3.3	18.1	305	.88?	?	38–127
119	163917	88048	17 59 01.5915	−09 46 25.075	46.8/152.69	G9III	2.39	21.9	536	1.28?	?	34–87

95	GJ777Ab				
157	GJ777Ac				
100		Tau Gruis			
101		Gamma Cephi			
103		Ogle-TR-56			
104		54 Piscis			
105		06 37 47.6189			
106		Ogle-TR-3			
108		BD+77 461			
109		CPC 19 552			
110		CD-48 2083			
111		CD-63 359			
112		CPC21.1 2334			
113		CPC20.1 4780			
114		CD-27 16109			
115		GJ 304			
116		CD-49 10033			
117		BD+17 1596			
118		91 Aqr			
		tau Gem			
		nu Oph			

	Star	Name	R.A.	Dec.	Distance	Type	Star	Planet	Period	Axis	Orbit	Lum
120		Ogle-TR-113b	10 52 24.40	-61 26 48.5	1,500/4,890	G	.79±.06	1.08±.28	1.443248	.02299±.00058	0	?
121		Ogle-TR-132b	10 50 34.72	-61 57 25.9	1,500/4,890	G	1.34±.10	1.19±.13	1.689857	.0306±.0008		?
122	37605	BD+05 985	05 40 01.7296	+06 03 38.085	42.88/139.79	K0V	.851	2.85±.26	54.2±.23	.26±.01	.736±.0095	.245-.525
123	26664	2M1207	12 07 33.4	-39 32 54	70/228	M8	25M_J	5±2	??	55AU	??	??
124		TrES-1	19 04 09.8	+36 37 57	157/511	K0V	.87±.03	.729±.036	3.030065	.0393±.0011	0	.18
126		Ogle-TR-111b	10 53 17.91	-61 24 20.3	850/2,657	G-K?	.82±.15 -.06	.53±.11	4.01610	.047±.001	0	.41
144		Ogle-TR-111c						.7±.2	16.0664±.005	.12±.01		
128	57087	GJ 436	11 42 11.0941	+26 42 23.652	10.23/33.357	M2.5	.42±.05	.067	2.644	.028	.12	.001-.025
129	88133	BD+18 2326	10 10 07.6754	+18 11 12.736	74.46/242.74	G5IV	1.2	.29	3.415±.001	.046	.11±.05	1.35-2.85
130	102117	CD-58 4207	11 44 50.4616	-58 42 13.354	42/136.92	G6V	.95±.05	.18±.03	20.8±.1	.15±.01	.08±.05	.95-1.11
131	117618b	CD-46 8708	13 32 25.5556	-47 16 16.906	38.02/123.95	G2V	1.05±.05	.16±.03	25.8±.5	.17±.01	.37±.1	99.166
						or —	.19±.04	52.2±.5	.28±.02	.39±.1		
132	208487	CD-38 14804	21 57 19.8477	-37 45 49.037	44/143.423	G2V	.95±.05	.45±.05	130±1	.49±.04	.32±.1	.42-1.72
133	154857b	CD-56 6717	17 11 15.7219	-56 40 50.865	68.54/223.44	G5V	1.17±.05	1.8	398.5	1.11	.5	2.77-5.05
134	154857c?							2?	3yrs?	1.9?	0	
136		Ogle-TR-10	17 51 28.25	-29 52 34.9	1500/4890	G2V or K	1.0±.05	.57±.12	3.10139	.04162±.00069	0	1.0
	202206b		21 14 57.7693	-20 47 21.154	46.33/151.03	G6V	1.15	17.42774	255.87±.06	.83040	.43492	.57-1.07
137	202206c							2.44	1,383.4±18.4	2.5420	.26692	
138	45350	BD-21 5972	06 28 45.7103	+38 57 46.667	40.94/159.56	G5V	1.02	.98	890.76±37.42	1.77	.78±.09	.73-1.44
139	99492	BD+39 1637	11 26 46.2771	+03 00 22.781	17.989/58.64	K2V	.78	.112	17.038	.119	.05±.12	.102-.259
140	117207	83 Leo B	13 29 21.1137	-35 34 15.589	33.01/107.62	G8 IV/V	1.04	2.06	2,627.08	3.78	.16±.08	.645-1.12
141	183263b	CD-34 8913	19 28 24.5727	+08 21 28.995	52.82/172.2	G2IV	1.17	3.69	634.23	1.52	.38±.03	.959-1.71
142	183263c	BD+08 4109						4 < M < 50	4-8yrs	> 2.5	?	
143	188015	BD+27 3539	19 52 04.5435	+28 06 01.356	52.63/171.5	G5IV	1.08	1.26	456.46	1.19	.15±.09	.64-1.22
145	93083	CD-32 7598	10 44 20.9149	-33 34 37.279	28.9/94.22	K2V/K3V	.70±.04	.37	143.58±.60	.477	.14±.03	.14-.342
146	102117	CPD-58 3823	11 44 50.4616	-58 42 13.354	30.49/99.39	K1V (G6V?)	.74±.05	.30	70.46±.18	.302	.11±.02	.784-1.551
147	142022	CD-83 202	16 10 15.0238	-84 13 53.802	35.87/90.88	G8/K0V	.99	4.4	1923±80	2.8	.57	.272-.588
148	2638	BD-06 82	00 29 59.872	-05 45 50.406	53.71/175.09	G5	.93	.48	3.4442±.0002	.044		.129-.413
149	27894	CD-59 829	04 20 47.0473	-59 24 39.014	42.37/138.135	K2V	.75	.62	17.991±.007	.122	.049±.008	.104-.262
150	63454	CD-77 298	07 39 21.8511	-78 16 44.30	35.80/116.72	K4V	.80	.38	2.81782±.000095	.036	0	.076-.190
151		GQ Lup	15 49 12.144	-35 39 03.9	140±50/456±163	K7V	.7	1-42	1,200yrs	103±37	?	?
152	13189		02 09 40.1717	+32 18 59.169	2,217-6,037ly	K2II	2-7	8-20	470.9±6	1.5-2.2	.28±.06	700-2750
153	8673		01 26 08.7849	+34 34 46.921	38.2/124.695	F7V	1.3	10-18	639	1.58	?	2.4-3.7
160	149026	BD+38 2787	16 30 29.6192	+38 20 50.315	78.9/257.098	G0 IV	1.3±.1	.36±.03	2.8766±.001	.042	?	
161	188753	BD+41 3535	19 54 58.3705	+41 52 17.511	44.823/146.12	G9V	1.06	1.114	3.35	.04522	?	1.8?

Pulsar Stars with Planets

Pulsar planets were discovered by measuring tiny pertubations in the timing of the radio signals received from the pulsars. Masses measured are in Earth (E) or Jupiter (J).

Star Name	R.A.	Dec.	Distance Parsecs/lyrs	Type	Star Sol=1	Planet Jup.=1	Period days	Axis (AU)	Orbit ecc.	Note
PSR 1257+12	13 00 02.99	+12 40 56.9	300/978	pulsar	?	.020 (E)	25.262	0.19	0.0	Not Applicable
						4.3 (E)	66.5419	.36	.0186	53° or 127° orbit inc.
						3.9 (E)	98.2114	.46	.0252	47° or 133° orbit inc.
						100 (E)	170 yrs	?	?	
PSR 1620−26	16 23 38.24	−26 31 53.9	3800/12,388	pulsar	1.35	2.5 (J)	100 yrs	23	?	55° orbital inclination

This pulsar has a White Dwarf Companion with a mass of .34, a period of 191 days, eccentricity of .025, and orbital inclination of 55 degrees, Lum = .003. The White Dwarf has an estimated age of 0.5 Gyrs.

References

Primary sources cited for discoveries

Abbreviations: AAS — American Astronomical Society; AJ — Astronomical Journal; ApJ — Astrophysical Journal; astro.ph. — Astrophysical Journal web archive for submitted papers.; A&A — Astronomy and Astrophysics; BAAS — Bulletin of the American Astronomical Society; ESA — European Space Agency; ESO — European Southern Observatory; GRA — Geophysical Research Abstracts; IAP — Institut D'Astrophysique de Paris; IAU — International Astronomical Union; IAUC — International Astronomical Union Circular; IJS — International Journal of Astrobiology; JGR — Journal of Geophysical Research; MNRAS — Monthly Notices of the Royal Astronomical Society; Nature — Nature magazine; PASP — Publications of the Astronomical Society of the Pacific; SIMBAD — Set of Identifications, Measurements, and Bibliography for Astronomical Data, maintained by the Centre de Données astronomiques de Strasbourg.

1. Latham, D., Stefanik, R., Mazeh, T., Mayor, M., Burki, G., 1989, Nature, 339, 38L.
2. Marcy, G. & Butler, R.P., 1997, ApJ, 481, 926.
3. Marcy, G. & Butler, R.P., 1996, ApJ, 464, L147.
4. Butler, R.P. & Marcy, G.W., 1996, ApJ, 464, L153.
5. Butler, R.P., Marcy, G.W., Williams, E., Hauser, H., Shirts, P., 1997, ApJ, 474, L115.
6. Gatewood, G., 1996, BAAS, 28, 885.
7. Cochran, W.D., Hatzes, A.P., Butler, R.P., Marcy, G.W., ApJ, 1997, 483, 457.
8. Noyes, R.W., Jha, S., Korzennik, S.G., Krockenberger, G., Nisenson, P., Brown, T., Kennelly, E.J., Horner, S., 1997, ApJ, 483, 111.
9. Marcy, G.W., Butler, R.P., Vogt, S.S., Fischer, D., Lissauer, J.L., 1998, ApJ, 505, L147.
10. Delfosse, X., Forveille, T, Mayor, M., Perrier, C., Naef, D., Queloz, D., 1998, A&A, 338, L67.
11. Butler, R.P., Marcy, G.W., Vogt, S., Apps, K., 1998, PASP, 110, 1389B.
12. Fischer, D.A., Marcy, G.W., Butler, R.P., Vogt, S.S., & Apps, K. 1999, PASP, 111, 50.
13. Queloz, D., Mayor, M., Weber, L., Blecha, A., Burnet, M., Confino, B., Naef, D., Pepe, F., Santos, N., Udry, S., 2000, A&A, 354, 99.
14. Els, S.G., Sterzik, M.F., Marchis, F., Pantin, E., Endl, M., Kürster, M., 2001, A&A 370, L1.
15. Leggett, S.K. TR Geballe, X. Fan, DP Schneider, JE Gunn et al., 2000, ApJ, 536, L35.
16. Marcy, G.W., Butler, R.P., Vogt, S.S., Liu, M.C., Laughlin, G., Apps, K., Graham, J.R., Lloyd, J., Luhman, K.L., Jayawardhana, R., 2001, ApJ, 555, 418.

17. Udry, S., Mayor, M., Naef, D., Pepe, F., Queloz, D., Santos, N.C., Burnet, M., Confino, B., Melo, C., 2000, A&A, 356, 590.
18. Marcy, G.W., Butler, R.P., Fischer, D.A., 1999, AAS, 194, 1402M.
19. Butler, R.P., Vogt, S.S., Marcy, G.W., Fischer, D.A., Brown, T.M., Contos, A.R., Korzennik, S.G., Nisenson, P., Noyes, R.W., 1999, ApJ, 526, 916.
20. Kuerster, M., Endl, M., Els, S., Hatzes, A.P., Cochran, W.D., Doebereiner, S., Dennerl, K., 2000, A&A, 353, L33.
21. Vogt, S.S., Marcy, G.W., Butler, R.P., Apps, K., 2000, ApJ, 536, 902.
22. Henry, G.W., Marcy, W., Butler, R.P., Paul, R., Steven, S., & Vogt, S.S. 2000, ApJ, 529, L41.
23. Naef, D., Mayor, M., Pepe, F., Queloz, D., Santos, N.C., Udry, S., Burnet, M., 2000, obswww.unige.ch/~udry/planet/gj3021_ann.html
24. Korzennik, S.G., Brown, T.M., Fischer, D.A., Nisenson, P., Noyes, R.W., 2000, ApJ, 533, L147.
25. Fischer, D.A., Marcy, G.W., Butler, R.P., Vogt, S.S., Frink, S., Apps, K., 2001, ApJ, 551, 1107.
26. Marcy, G.W., Butler, R.P., Vogt, S.S. 2000, ApJ, 536, L43.
27. Butler, R.P., Vogt, S.S., Marcy, G.W., Fischer, D.A., Henry, G.W., Apps, K. 2000, ApJ, 545, 504.
28. ESO. 2000, ESO Press Release: Exoplanet galore!, http://www.eso.org/outreach/pressrel/pr2000/pr13-00.html.
29. Mayor, M., Naef, D., Pepe, F., Queloz, D., Santos, D., Santos, N.C., Udry, S., Burnet, M., Perrier-Bellet, C., Beuzit, J.L., Sivan, J.P., 2000, IAU press release, Aug. 7th.
30. Hatzes, A., Cochran, W., McArthur, B., Baliunas, S.L., Walker, G., et al., 2000, ApJ, 544, L145.
31. Tinney, C.G., Butler, R.P., Marcy, G.W., Jones, H.R.A., Penny, A.J., Vogt, S.S., Apps, K., Henry, G.W., 2001, ApJ, 551, 507.
32. Butler, R.P., Tinney, C.G., Marcy, G.W., Jones, H.R.A., Penny, A.J., Apps, K., 2001, ApJ, 555, 410.
33. Marcy, G.W., Butler, R.P., Fischer, D., Vogt, S.S., Lissauer, J.L., Rivera, E.J., 2001, ApJ, 556, 296.
34. ESO. 2001, ESO Press Release: Exoplanets: The Hunt Continues!, http://www.eso.org/outreach/pressrel/pr-2001/pr0701.html.
35. Fischer, D.A., Marcy, G.W., Butler, R.P., Laughlin, G., Vogt, S.S., 2002, ApJ, 564, 1028.
36. Vogt, S.S., Butler, R.P., Marcy, G.W., Fischer, D.A., Pourbaix, D., Apps, K., Laughlin, G., 2002, ApJ, 568, 352.
37. Tinney, C.G., Butler, R.P., Marcy, G.W., Jones, H.R.A., Penny, A.J., McCarthy, C., Carter, B.D., 2002, ApJ, 571, 528.
38. Jones, H.R.A., Butler, R.P., Tinney, C.G., Marcy, G.W., Penny, A.J., McCarthy, C., Carter, B.D., Pourbaix, D., 2002, MNRAS, 333, 871.
39. Frink, S., Mitchell, D.S., Quirrenbach, A., Fischer, D.A., Marcy, G.W., Butler, R.P., 2001, AAS, 199, 6104.
40. Fischer, D.A., Marcy, G.W., Butler, R.P., Vogt, S.S., Walp, B., Apps, K., 2002, PASP, 114, 529.
41. Marcy, G.W., Butler, R.P., Fischer, D., Laughlin, G., Vogy, S.S., Henry, G.W., Pourbaix, D., 2002, ApJ, 581, 1375.
42. Fischer, D.A., Marcy, G.W., Butler, R.P., Vogt, S.S., Henry, G.W. ., Pourbaix, D., Walp, B., Misch, A.A., Wright, J., 2003, ApJ, 586, 1394.
43. Butler, R.P., Marcy, G.W., Vogt, S.S., Fischer, Henry, G.W., Laughlin, G., Wright, J.T., 2003, ApJ, 582, 455.
44. Tinney, C.G., Butler, R.P., Marcy, G.W., Jones, H.R.A., Penny, A.J., McCarthy, C., Carter, B.D., Bond, J., 2002, ApJ, 587, 423.
45. Jones, H.R.A., Butler, R.P., Marcy, W., Tinney, C.G., Penny, A.J., McCarthy, C., Carter, B.D., 2002, MNRAS, 337, 1170.
46. Perrier, C., "Scientific Frontiers in Research on Extrasolar Planets" conference, Washington D.C., June, 19th 2002.
47. Jones, H.R.A., Butler, R.P., Marcy, W., Tinney, C.G., Penny, A.J., McCarthy, C., Carter, B.D., 2003, MNRAS, 341, 948.
48. Cochran, W.D., Hatzes, A.P., Endl, M., Paulson, D.B., Walker, G.A.H., Campbell, B., Yang, S., 2002, DPS 34th Meeting. BAAS, 34, 42.02
49. Konacki, M., Torres, G., Jha S., Sasselov D., 2003, Nature, 421, 507.
50. Fischer, D.A., Butler, R.P., Marcy, G.W., Vogt, S.S., Henry, G.W., 2003, ApJ, 590, 1081.
51. ESO 2003, ESO Press Release: Distant World in Peril Discovered From La Silla, http://www.eso.org/outreach/pressrel/pr-2003/pr-03-03.html.
52. Dreizler, S., Hauschildt, P., Kley, W., Rauch, T., Schuh, S.L., Werner, K., Wolff, B., 2003, A&A, 402, 791.
53. Udry, S., M. Mayor, Clausen, J.V.,

Freyhammer, L.M., Helt, B.E., Lovis, C., Naef, D., Olsen, E., Pepe, F., Queloz, D., Santos, N.C., 2003, A&A 407, 679.

54. Sato, B., Ando, H., Kambe, E., Takeda, Y., Izumiura, H., Masuda, S., Watanabe, E., Noguchi, K., Wada, S., Okada, N., Koyano, H., Maehara, H., Norimoto, Y., Okada, T., Shimizu, Y., Uraguchi, F., Yanagisawa, K., Yoshida, M., 2003, ApJ, 597, L157.

55. IAP Colloquium, XIX, 2003, "Extrasolar planets: Today & Tormorrow," 7 planets detected with CORALIE.

56. Carter, B.D., Butler,R.P., Tinney, C.G., Jones, H.R.A., Marcy, G.W., McCarthy, C., Debra A. Fischer, D.A., Penny, A.J., 2003, ApJ, 593, L43.

57. Mayor, M., Pepe, F., Queloz, D., Bouchy, F., Rupprecht, G., Curto, G.L., Avila, G.,Benz, W., Bertaux, J.-L., Bonfils, X., Dall, Th, Dekker, H., Delabre, B., Eckert, W., Fleury, M., Gilliotte, A., Gojak, D., Guzman, J.C., Kohler, S., Lizon, J.-L., Longinotti, A., Lovis, C., Megevand, D., Pasquini, L., Reyes, J., Sivan, J.-P., Sosnowska, D., Soto, R., Udry, S., Van Kesteren, A., Weber, L., Weilenmann, U., 2003, ESO Messenger, 114, 21.

58. Mitchell, D.S., Frink, S., Quirrenbach, A., Fischer, D.A., Marcy, G.W., Butler, R.P., AAS 203rd Meeting, 2003, BAAS, 35#5.

59. Boucher, F., Pont, F., Santos, N.C., Melo, C., Mayor, M., Queloz, D., Udry, S., 2004, A&A, 421,L13.

60. Cochran, W.D., Endl, M., McArthur, B., Paulson, D.B., Smith, V.V., MacQueen, P.J., Tull, R.G., Good, J., Booth, J., Shetrone, M., Roman, B., Odewahn, S., Deglman, F., Graver, M., Soukup, M., Villarreal Jr., M.L., 2004, ApJ, 611, Letters.

61. Chauvin, G., Lagrange, A.-M., Dumas, C., Zuckerman, B., Mouillet, D., Song, I., Beuzit, J.-L., Lowrance, P., 2004, A&A, 425, L29.

62. Alonso, R., Brown, T.M., Torres, G., Latham, D.W., Sozzetti, A., Mandushev, G., Belmonte, J.A., Charbonneau. D., Deeg, H.J., Dunham, E.W., O'Donovan, F.T., Stefanik, R.P., 2004, ApJ, Letters, 613, L153.

63. Santos, N.C., Bouchy, F., Mayor, M., Pepe, F., Queloz, D., Udry, S., Lovis, C., Bazot, M., Benz, W., Bertaux, J.-L., Lo Curto, G., Delfosse, X., Mordasini, C., Naef, D., Sivan, J.-P., Vauclair, S., 2004, A&A, 426, L19.

64. Pont, F., Bouchy, F., Queloz, D., Santos, N.C., Melo, C., Mayor, M., Udry, S., 2004, A&A, 426, L15.

65. McArthur, B.E., End, M., Cochran, W.D., Benedict, G.F., Fischer, D.A., Marcy, G.W., Butler, R.P., Naef, D., Mayor, M., Queloz, D., Udry, S., 2004, ApJ, 614, L81.

66. Butler, R.P., Vogt, S.S., Marcy, G.W., Fischer, D.A., Wright, J.T., Henry, G.W., Laughlin, G., Lissauer, J.J., 2004, ApJ, 617, 580.

67. Fischer, D.A., Laughlin, G., Butler, R. P., Marcy, G.W., Johnson, J., Henry, G., Valenti, J., Vogt, S.S., et al., 2005, ApJ, 620, 481.

68. Tinney, C.G., Butler, R.P., Marcy, G.W., Jones, H.R.A., Penny, A.J., McCarthy, C., Carter, B.D., Fischer, D.A., 2004, ApJ, 623,1171.

69. McCarthy, C., Butler, R.P., Tinney, C.G., Jones, H.R.A., Marcy, G.W., Carter, B., Penny, A.J., Fischer, D.A., 2004, ApJ, 617, 575.

70. Bouchy, F., Pont, F., Santos, N., Melo, C., Mayor, M., Queloz, D., Udry, S., 2005, A&A, 421, L13.

71. Correia, A.C.M., Udry, S., Mayor, M., Laskar, J., Naef, D., Pepe, F., Queloz, D., Santos, N.C., 2005, A&A, submitted.

72. Marcy, G.W., Butler, R.P., Vogt, S.S., Fischer, D.A., Henry, G.W., Laughlin, G., Wright, J.T., Johnson, J.A., 2005, ApJ, 619, 570.

73. Minniti, D., 2005, A&A, accepted.

74. Lovis, C., Mayor, M., Pepe, F., Queloz, D., Santos, N., Sosnowska, D., Udry, S., Benz, W., Bertaux, J.-L., Bouchy, F., Mordasini, C., Sivan, J.-P., 2005, A&A, submitted.

75. Moutou, C., Mayor, M., Bouchy, F., Lovis, C., Pepe, F., Queloz, D., Santos, N., Sosnowska, D., Udry, S., Benz, W., Naef, D., Segresan, D., & Sivan, J.-P., 2005, A&A, submitted.

76. Aspen Winter Conference 2005 "Planet Formation and Detection," Aspen, February 8, 2005.

77. Neuhauser, R., Guenther, E., Wuchterl, G., Mugrauer, M., Bedalov, A., Hauschildt, P., 2005, A&A, accepted.

78. Hatzes, Guenther, E., 2005, GRA, Vol 7, 06592.

79. Menou, K., Serge Tabachnik, S., 2003, ApJ, 583, 473.

80. Asghari, N., Broeg, C., Carone, L., Casas-Miranda, R., 2004, A&A, 426, 353.

81. Dvorak, R., Pilat-Lohinger, E., Funk, B., Freistetter, F., 2003, A&A, 398, L1.

82. Leger, A., Selsis, F., Sotin, C., Guillot, D., Despois, D., Mawet, D., Ollivier, M.,

Labeque, A., Valette, C., Brachet, F., Chazalas, B., Lammer, H., 2004, Icarus Notes.

83. Desidera, S., Gratton, R.G., Endl, M., Barbieri, M., Claudi, R.U., Cosentino, R., Lucatello, S., Marzari, F., Scuderi, S., 2003, A&A, 405, 207.

84. Ibid, 2004, A&A, 420, L27.

85. James B. Kaler, Cambridge University, 1989, ISBN 0–521–30494–6.

86. Underwood, D.R., Jones, B.W., Sleep, P.N., 2003, accepted by IJS, 2003astro.ph.12522U.

87. Wolszczan, A., Frail, D., 1992, Nature, 255, 145.

88. Rasio, F.A., 1994, ApJ, 427, 107L.

89. Dvorak, R., Pilat-Lohinger, E., Schwarz, R., Freistetter, F., 2004, A&A, 426, L37.

90. Barnes, J.W., O'Brien, D.P., 2002, ApJ, 575, 1087.

91. Osorio, M.R.Z., et, al., 2002, Science, 290, 103.

92. Burgasser, A.J., Kirkpatrick, J.D., McGovern, M.R., McLean, I.S., Prato, L., Reid, I.N., 2004, ApJ, 604, 827.

93. Bond, I.A., Udalski, A., Jaroszynski, M, et. al., 2004, 606, L000.

94. Rice, W.K.M., Wood, K., Armitage, P.J., Whitney, B.A., Bjorkman, J.E., 2003, MNRAS, 342, 79.

95. Gonzalez, G., Brownlee, D., Ward, P.D., 2001, Icarus, 152, 1, 185.

96. Brown, T.M., Charbonneau, D., Gilliland, R.L., Albrow, M.D., et al., 2000, AAS, 196th meeting 2.03.

97. Zinnecker, H, 2003, astro.ph., 1080. (eprint)

98. Rivera, E.J., Lissauer, J.J.Butler, P., Marcy, G. Vogt, S., Fischer, D., Brown, T., Laughlin, G., 2005, ApJ, Letters, submitted.

99. Vogt., S., Butler, P., Marcy, G., Fischer, D., Henry, G., Laughlin, G., Wright, J., Johnson, A., 2005, ApJ, accepted.

100. Sato, B., Fischer, D., Henry, G., Laughlin, G., Butler, R., Marcy, G.W., Vogt, S., Bodenheimer, P., Ida, S., Toyota, E., Wolf, A., Valenti, J., Boyd, L., Johnson, J., Wright, J., Ammons, M., Robinson, S., Strader, J., McCarthy, C., Tah, K., Minniti, D., 2005, ApJ, accepted.

101. Konacki, M, 2005, Nature, 436, 230.

Bibliography

Supplemental sources used to update information from original papers

Abbreviations: AAS — American Astronomical Society; AJ — Astronomical Journal; ApJ — Astrophysical Journal; astro.ph. — Astrophysical Journal web archive for submitted papers.; A&A — Astronomy and Astrophysics; BAAS — Bulletin of the American Astronomical Society; ESA — European Space Agency; ESO — European Southern Observatory; GRA — Geophysical Research Abstracts; IAP — Institut D'Astrophysique de Paris; IAU — International Astronomical Union; IAUC — International Astronomical Union Circular; IJS — International Journal of Astro-biology; JGR — Journal of Geophysical Research; MNRAS — Monthly Notices of the Royal Astronomical Society; Nature — Nature magazine; PASP — Publications of the Astronomical Society of the Pacific; SIMBAD — Set of Identifications, Measurements, and Bibliography for Astronomical Data, maintained by the Centre de Données astronomiques de Strasbourg.

Baraffe, I., Chabrier, G., Barman, T.S., Allard, F., Hauschildt, P.H., 2003, A&A 402, 701.
Barbieri, M., Gratton, R.G., 2002, ApJ, 384, 879.
Barge, P. & Viton, M., 2003, ApJ, 593, L117.
Barnes, R., Quinn, T., 2004, ApJ., submitted.
Barnes, R., Raymond, S., 2004, asttro.ph., 2542, submitted Feb. 2004.
Benedict, G.F., McArthur, B., STScI-2002-27 (Hubble Telescope).
Bennett, D.P., Rhie, S.H., Becker, A.C., et. al., 1999, Nature, 402, 57.
Burgasser, A.J., Kirkpatrick, J.D., McGovern, M.R., McLean, I.S., Prato, L., Reid, I.N., 2004, ApJ, 604, 827.
Butler, R.P., Marcy, G.W., Vogt, S.S., Fischer, D.A., Brown, T.M., Contos, A.R., Korzennik, S.G., Nisenson, P., Noyes, R.W., 1999, ApJ, 526, 916.
Butler, R., Marcy, G., Vogt, S., Tinney, C., Jones, H.R.A., McCarthy C., Penny A., Apps K., Carter B., 2002, ApJ, 578, 565.
California & Carnegie Planet Search website, exoplanets.org/.
Cameron, C.A., Horne, K., Penny, A., James, D., 1999, Nature 402, 751.
Cameron, C.A., Horne, K. James, D., Penny A., Semel, M., 2000, Planetary Systems in the Universe, IAU Symp 202.

Bibliography

Chabrier, G., Barman, T., Baraffe, I., Allard, F., Hauschildt, P., 2004, ApJ. Letters, accepted.
Charbonneau, D., Allen, L., Megeath, T., Torres, G., Alonso, R., Brown, T., Gilliland, R., Latham, D.W., Mandushev, G., O'Donovan, F.T., Sozzetti, A., 2005, ApJ, 20 June.
Chen, Y.Q., Zhao, G., 2001, A&A, 374, L1.
Cho, J.Y-K., Menou, K., Hansen, B., Seager, S., 2003, ApJ, 587, L117.
Cochran, W. D., Hatzes, A. P. & Paulson, D. B. 2002, AJ, 124, 565.
Cody, A.M., Sasselov, D.D., 2001, ApJ, 568, 377.
Debes, J.H., Steinn, S., 2002, ApJ, 572, 556.
Deming, D., Seager, S., Richardson, L.J., Harrington, J., 2005, Nature, in press.
Drake, A., Cook, K., 2004, ApJ, 604, 379.
Dreizler, S., Rauch, T., Hauschildt, P., Schuh, S.L., Kley, W., Werner, K., 2002, A&A, 391, L17.
Ecuvillon, A., Israelian, G., Santos, N.C., Mayor, M., Garcia Lopez, R.J., Randich, S., 2004, A&A, 418, 703.
Eggenberger, A., Udry, S., Mayor, M., 2004, A&A, 417, 353.
ESA. 1997, The HIPPARCOS and TYCHO catalogue, ESA-SP 1200.
Extrasolar Planets Encyclopaedia website, www.obspm.fr/encycl/encycl.html.
Fernandez, J., Santos, N.C., 2004, A&A, 427, 607.
Fischer, D.A., Marcy, G.W., Butler, R.P., 2001, http://expplanets.org/esp/47uma/47uma_announce.html.
Franck, S., von Bloh, W., Bounama, C., Steffen, M., Schonberner, D., Schellnhuber, H.-J., 2000, JGR, 105, E1, 1651.
Frink, S., Mitchell, D.S., Quirrenbach, A., Fischer, D.A., Marcy, G.W., Butler, R.P., 2002, ApJ 576, 478, 2002.
Geneva Extrasolar Planet Search Programmes, The, website, obswww.unige.ch/~udry/planet/planet.html.
Gonzalez, G., 1996, MNRAS, 285, 203.
Gonzalez, G., Laws, C., 2000, AJ, 119, 390.
Gozdziewski, K., Konacki, M., Wolszczan, 2005, ApJ, 619, 1084.
Hatzes, A., Cochran, W., Bakker, E.J., 1998, Nature, 391, 154.
Hauser, H.M., Marcy, G.W., 1999, PASP, 111, 321H.
Henry, G.W., Marcy, G., Butler, R.P., Vogt, S.S., 1999, IAUC, 7307.
Israelian, G., Santos, N.C., Mayor, M., Rebolo, R., 2001, Natur, 411, 163.
Ivanov, P.B., Paploizou, J.C.B., 2004, MNRAS, 347, 437
Jianghui, J., Kinoshita, H., Lin, L., Guangyu, L., Nakai, H., 2002, ApJ, submitted, 2002astro.ph. .8025J.
Jones, B.W., Sleep, P.N., 2002, A&A, 393, 1015.
Jorissen, A., Mayor, M., Udry, S., 2001, A&A, 379, 992.
Kiseleva-Eggleton, L., Bois, E., Rambaux, N., Dvorak, R., 2002, ApJ, 578, L145.
Konacki, M., Torres, G., Jha S., Sasselov D., 2004, ApJ, 609, L37.
_____. 2004, ApJ, accepted Dec.
Laughlin, G. Chambers, J., Fischer, D., 2002, ApJ, 579, 455.
Laughlin, G. Chambers, J., 2002, AJ, 124, 592.
Laughlin, G., Wolf, A., Vanmunster, T., Bodenheimer, P., Fischer, D., Marcy, G., Butler, P., Vogt, S., 2004, ApJ, accepted.
Lowrance1, P., Becklin, E.E., Schneider, S., 2001, ASP Conf. series.

Laws, Ch., Gonzalez, G., Walker, K., Tyagi, S., Dodsworth, J., Snider, K., Sunitzeff, N., 2003, A.J., accepted
Lee, M.H., Peale, S.J., 2002, ApJ, 567, 596.
Lowrance, P.J., Kirkpatrick, J.D., Beichman, C.A., 2002, ApJ, 572, L79.
Marcy, G.W., Butler, R.P., Vogt, S.S., Fischer, D., Liu, M.C., 1999, ApJ, 520, 239.
Martin, E.L., Osorio, M.R.Z., 2003, ApJ, 593, L113.
Mayor, M., Udry, S., Naef, D., Pepe, F., Queloz, D., Santos, N., Burnet, M., 2004, A&A, 415, 391.
McGrath, M.A., Nelan, E., Black, D.C., Gatewood, G., Noll, K., Schultz, A., Lubrow, S., Han, I., Stepinski, T.F., Targett, T., 2002, ApJ, 564, L27.
Moutou, C., Pont, F., Bouchy, F., Mayor, M., 2004, A&A, 424, L31.
Mugrauer, M., Neuhauser, R., Mazeh, T., Alves, J., Guenther, E., 2004, A&A, 425, 249.
Naef, D., Mayor, M., Korzennik, S.G., Queloz, D., Udry, S., Nisenson, P., Noyes, R.W., Brown, T.M., Beuzit, J.L., Perrier, C., Sivan, J.P., 2003, A&A, 410, 1051.
Naef, D., Latham, D.W., Mayor, M., Mazeh, J.L., Beuzit, J.L., Drujier, G.A., Perrier-Bellet, C., Queloz, D., Sivan, J.P., Torres, G., Udry, S., Zucker, S., 2001, A&A, 375, L27.
Naef, D., Mayor, M., Pepe, F., Queloz, D., Santos, N.C., Udry, S., Burnet, M., 2001, A&A, 375, 205.
Naef, D., Mayor, M., Beuzit, J.L., Perrier, C., Queloz, D., Sivan, J.P., Udrys, S., 2004, A&A, 414, 351.
Noble, M., Musielak, Z.E., Cuntz, M., 2002, ApJ, 572, 1024.
Pepe, F., Mayor, M., Galland, F., Naef, D., Queloz, D., Santos, N.C., Udry, S., Burnet, M., 2002, A&A, 388, 632.
Perrier, C., Sivan, J.-P., Naef, D., Beuzit, J.L., Mayor, M., Queloz, D., Udry, S., 2003, A&A, 410, 1039.
Patience, J., White, R.J., Ghez, A.M., McCabe, C., McLean, I.S., Larkin, J.E., Prato, L., Kim, S.S., Lloyd, J.P., Liu, M.C., Graham, J.R., Macintosh, B.A., Gavel, D.T., Max, C.E., Bauman, B.J., Olivier, S.S., Wizinowich, P., Acton, D.S., 2002, ApJ., 581, 654.
Queloz, D., Eggenberger, A., Mayor, M., PerrierC., Beuzit, J.L., Naef, D., Sivan, J.P., Udry, S., 2000, A&A, 359, L13.
Quillen, A.C., Thorndike, S., 2002, ApJ, 578, L149.
Raymond, S.N., Barnes, R., 2004, astro.ph., 4212R.
Rivera, E.J., Lissauer, J.J., 2001, ApJ, 558, 392.
Santos, N.C., Israelian, G., Mayor, M. 2004, A&A, 415, 1153.
Santos, N.C., Israelian, G., Mayor, M., Rebolo, R., Udry, S., 2003, A&A, 398, 363.
Santos, N.C., Mayor, M., Naef, D., Pepe, F., Queloz, D., Udry, S., Burnet, M., Clausen, J.V., Helt, B.E., Olsen, E.H., Pritchard, J.D., 2002, A&A, 392, 215.
Santos, N.C., Mayor, M., Naef, D., Pepe, F., Queloz, D., Udry, S., & Burnet, M. 2001, A&A, 379, 999.
Santos, N.C., Udry, S., Mayor, M., Naef, D., Pepe, F., Queloz, D., Burki, G., Cramer, N., Nicolet, B., 2003, A&A, 406, 373.
Santos, N.C., Israelian, G., Mayor, M. 2000, A&A, 363, 228.
Setiawan, J., Hatzes, A.P., von der Lühe, O., Pasquini, L., Naef, D., da Silva, L., Udry, S., Queloz, D., Girardi, 2003, A&A, 398, L19.
Smith, V.V., Cunha, K., Lazzaro, D. 2001, AJ, 121, 3207.

Soubiran, C., Triaud, A., 2004, A&A, 418, 1089.
Udry, S., Mayor, M., Naef, D., Pepe, F., Queloz, D., Santos, N. C. Burnet, M., 2002, A&A, 390, 267.
Udry, S., Mayor, M., Naef, D., Pepe, F., Queloz, D., Santos, N.C., Burnet, M., Confino, B., Melo, C., 2000, A&A, 356, 590.
Udry, S., Mayor, M., Naef, D., Pepe, F., Queloz, D., Santos, N.C., Burnet, M., Confino, B., Melo, C., 2003c, A&A, 407, 679.
Villard, R., 2003, Astronomers unravel the strange life and hard times of the farthest and oldest known planet, Http:\\currents.usc.edu/03-04/07-21/planet.html
Zucker, S., Mazeh, T., 2001, ApJ, 562, 549.
Zucker, S., Mazeh, T., Santos, N.C., Mayor, M., 2004, A&A, 426, 695.
Zucker, S., Naef, D., Latham, D.W., Mayor, M., Mazeh, T., Beuzit, J.L., Drukier, G., Perrier-Bellet, C., Queloz, D., Sivan, J.P., Torres, G., Udry, S., 2002, ApJL, 568, 363.

Index

ADS 31
ADS 13886B 50
ADS 16642 171
Aitken, R.G. 31
Aldebaran 22, 23
Alpha Bootis 7
Alpha Centauri 24, 32, 42
Altair 22
Andromeda 43, 59, 168
Andromeda galaxy 32
Antares 22, 23
Antila 160
Apastron 64
Apparent Magnitude 34
Aquarius 46, 49, 63, 140, 148
Aquila 57, 158
Ara 66, 85
Arcturus 7, 22, 23
Aries 114
Astronomical Unit 33
Astronomy 11, 14
The Astronomy and Astrophysical Journal 8
The Astrophysical Journal 8
Auriga 105, 124, 155, 175
BAAS 187

Barnard's Star 22, 42
BD -02 5917 171
BD -10 3166 70
BD +18 4505C 50
BD +59 1655 103
Betelgeuse 22, 23
Bode's Law 15
Bonner Durchmusterung 31
Bootes 41, 109
Bright Star Catalog 69

Brown dwarf 21, 154

California & Carniegie Planet Search website 8, 180
Callisto 14
Camelopardalis 130
Cancer 40
Canopus 23
Cape Photographic Durchmusterung 31
Capella 23
Capricorn 154
Carinae 133, 141, 142, 146
Celsius 11
Centaurus 80, 108, 118, 144, 149, 150, 157, 161
Centigrade 11, 33
Cephi 123
Cetus 67, 79, 162
Chamaeleon 164
Chromosphere 34
Coma Berenices 36, 107
Cordoba Durchmusterung 31
Corona 169, 170
Corona Borealis 45
Coyote 39
CPD 31
Crater 69, 70
Crux 73
Cygnus 44, 48, 117, 170

Darwin, Charles 16
Darwin, George 16
Declination 34
Delphinus 50
Delta Orionis 23
Deneb 22

195

Index

Dione 29
Draco 103
Dust Rings 175
Dwarf 23

Eccentricity 33
83 Leo B 156
Elrai 123
Epimetheus 16
Epsilon Eridani 22, 82
Epsilon Reticuli 84
Eridanus 52, 82, 87, 131
Eta Carinae 14
Europa 14
Extra-solar Planets Catalog 8, 180

Fahrenheit 11, 33
51 Pegasi 6, 37
54 Piscis 126
55 Cancri 40
Fomalhaut 175
47 Tucanae 26
47 Ursae Majoris 39
14 Hercules 22, 47

Galactic Bulge 174
Gamma Cephi 123
Ganymede 14
Gemini 88, 139
Gl 777Ba 117
Gl 777Bb 117
Gliese 86 52
Gliese 229B 23
Gliese 476 147
Gliese 614 47
Gliese 691 85
Gliese 777A 117
Gliese 876 46, 147
Gleise 3021 64
GM Aurigae 175
GQ Lupus 166
Greater Magellanic Cloud 26
Grus 94, 122, 151, 152
Gurps Space 8

HD 31
HD 142A 100
HD 1237 64
HD 2039 120
HD 2638 162
HD 3651 126

HD 4203 95
HD 4208 96
HD 6434 78
HD 8574 86
HD 8673 168
HD 9826 43
HD 10647 131
HD 10697 59
HD 12661 66
HD 13189 167
HD 13445 52
HD 16141 67
HD 17051 55
HD 19994 79
HD 20367 114
HD 22049 82
HD 23079 101
HD 23596 115
HD 27442 84
HD 27894 163
HD 28185 87
HD 30177 110
HD 33636 97
HD 37124 60
HD 37605 143
HD 38529 76
HD 39091 102
HD 40979 124
HD 41004A 132
HD 45350 155
HD 46375 68
HD 47536 127
HD 49674 105
HD 50554 88
HD 52265 69
HD 53911 175
HD 59686 139
HD 63454 164
HD 65216 133
HD 68988 98
HD 70642 137
HD 72659 106
HD 73256 129
HD 73526 111
HD 74156 89
HD 75289 54
HD 75732 40
HD 76700 121
HD 80606 90
HD 80607 90
HD 82943 71

Index

HD 83443 72
HD 88133 148
HD 89744 65
HD 92788 77
HD 93083 160
HD 95128 39
HD 95735 42
HD 99492 156
HD 101930 161
HD 102117 149
HD 104985 130
HD 106252 91
HD 108147 73
HD 108874 107
HD 111232 22, 134
HD 114386 118
HD 114729 108
HD 114762 5, 31, 36
HD 114783 22, 99
HD 117176 38
HD 117207 157
HD 117618 150
HD 120136 41
HD 121504 80
HD 128311 109
HD 130322 56
HD 134987 61
HD 136118 104
HD 137759 103
HD 141937 92
HD 142022A 165
HD 142415 135
HD 143761 17, 45
HD 145675 47
HD 147513 119
HD 149026 169
HD 150706 116
HD 154857 152
HD 160691 85
HD 168443 53
HD 168746 74
HD 169830 75
HD 177830 62
HD 178911A 93
HD 178911Aa 93
HD 178911Ab 93
HD 178911B 93, 171
HD 179949 83
HD 183263 158
HD 186427 44
HD 187123 48

HD 188015 159
HD 188753 170
HD 190228 81
HD 192263 57
HD 195019 50
HD 196050 112
HD 196360 117
HD 202206 154
HD 208487 151
HD 209458 6, 31, 58, 128, 141, 142, 153
HD 210277 49
HD 213240 94
HD 216435 122
HD 216437 113
HD 216770 136
HD 216956 175
HD 217014 37
HD 217107 51
HD 219449 140
HD 219542A 171
HD 219542B 171
HD 222404 123
HD 222582 63
HD 330075 138
Henry-Draper 31
Hercules 47
Hertzsprung-Russell Diagram 23
HIP 40952 137
HIP 57087 147
HIP 114914 171
Hipparcos 66, 134, 149
Horologium 55, 101
HR 810 55
HR 8728 175
Hubble Space Telescope 11
Hydra 71, 89, 106, 175
Hydrus 64

Indus 113
Io 14, 146
Iota Draconis 103
Iota Horologii 55

Janus 16

Kelvin 11, 33

Lagrange, Louis 29
Lagrange Points 29, 96, 99, 101
Lalande 21185 22, 42

Latham, D.W. 5
Leo 147, 148, 156
Lesser Magellanic Cloud 26
Libra 61, 92
Luminosity 33
Lupus 166
Lyra 62, 93
Lyre 145

M4 27
MACHO-97-BLG-41 174
MACHO Project 173
Magellanic Clouds 26, 174
Mars 20
Mensa 102
Metallicity 34
Microlensing 174
Microwave Anisotropy Probe 29
Milky Way galaxy 26, 173
Mintaka 22
Mira 22
MOA 2003-BLG-53 174
Monoceros 68
Mu Arae 85
Musca 134

91 Aqr 140
Norma 138

Octans 165
OGLE 174
OGLE-TR-3 128
OGLE-TR-10 153
OGLE-TR-56 125, 153
OGLE-TR-111 146
OGLE-TR-113 141
OGLE-TR-132 142
OGLE 2003-BLG-235 174
1RXSJ155740.7-601154 135
Oort-cloud 25
Optical Gravitational Lensing Experiment 174
Orion 76, 97, 143, 176
Orion Arm 125, 153

Pavo 112
Pegasus 37, 58
Periastron 64
Perseus 115
Phoenix 78, 100, 120
Pi Mensae 102

Pictor 132
Pisces 51, 86, 95, 126, 127
Piscis Austrinus 136, 175
Pleiades 11
Polaris 22
Population I 26
Population II 24
Procyon A 22, 23
Procyon B 22
PSR 1257+12 27, 186
PSR 1620-26 27, 186
Pulsar Planets 26
Pulsar Stars 186
Puppis 137
Pyxis 129

Q0957+561 175

Reticulum 84, 110, 163
Rho Cancri 40
Rho Coronae Borealis 17, 45
Rigel 22, 23
Right Ascension 34
Rosat-All-Sky-Survey 135
Ross 780 22, 46, 147
Ross 905 147
Rossiter, R.A. 31
Royal Greenwich Observatory 31

S Orionis 70 12, 176
Sagittarius 75, 83, 125, 153, 174
Sagittarius Arm 125, 153
Saturn 16
Scorpius 119, 128
Sculptor 96
Scutum 74, 174
Semi-major axis 34
Serpens 53, 104
Serpens Caput 104
70 Virginis 38
79 Ceti 67
Sextan 77
Sigma Orionis cluster 12
SIMBAD 8, 66, 69, 108, 109, 149, 160, 180, 187
Sirius 32
Sirius A 22, 23
Sirius B 22
16 Cyg A 44
16 Cyg B 171
16 Cygni A 44

Index

16 Cygni B 44
Sky & Telescope 16
Solar and Heliospheric Observatory 29
Spica 23
Stars and Their Spectra 8
Steele, Allen 39
Stellar wind 28
Steve Jackson Games 8
Supergiant 23

T Tauri star 166, 175
Tau Bootis 22, 41
Tau Gruis 122
Taurus 60
10 Lacertra Mintaka 23
Tethys 29
Titan 14, 20
Transit Detection 173
TrES-1 145
Triangulum 167
Triangulum Australe 135

Triton 14
Tucanae 26
TW Hydrae 6 175
TW Hydrae Association 144
2M1207 144

Upsilon Andromeda 22
Upsilon Andromedae 43
Ursa Major 39, 42, 65, 90, 98, 119, 175
Ursa Minor 116

Vega 22, 23
Vela 54, 72, 111
Venus 20
Virgo 38, 56, 91, 99
Volans 121
Vulpecula 81, 159, 160

White dwarf 21, 23, 119
Wolf, Professor 31
Wolf 359 42